FINDING
SAMUEL LOWE

FINDING SAMUEL LOWE

CHINA, JAMAICA, HARLEM

PAULA WILLIAMS MADISON

Amistad

An Imprint of HarperCollins*Publishers*

HarperCollins books may be purchased for educational, business, or sales promotional use. For information, please e-mail the Special Markets Department at SPsales@harpercollins.com.

FIRST EDITION

Unless otherwise indicated, all photographs are courtesy of the Williams Family Photo Collection.

Designed by Suet Yee Chong

Library of Congress Cataloging-in-Publication Data has been applied for.

ISBN: 978-0-06-233163-2

15 16 17 18 19 OV/RRD 10 9 8 7 6 5 4 3 2 1

In this country American means white.
Everybody else has to hyphenate.

—TONI MORRISON, *BELOVED*

NELL VERA LOWE WILLIAMS—BI SHAN LOWE

JAMAICA, 1940s

TO MY GRANDFATHER
Low Ding Chow
— Samuel Lowe —

TO MY MOTHER
Nell Vera Lowe Williams—
— Bi Shan Lowe —

Found.
Claimed.
Loved.

CONTENTS

FINDING
SAMUEL LOWE

PROLOGUE

E GAVE ME ONE OF HIS SERIOUS, TWENTY-FOUR-KARAT, ALL-ROOSEVELT looks. "Baby," he asked, "do you know you're Black?"

I looked at him, puzzled and defensive. "Yes. I know I am Black."

He looked a little hesitant, as if he were afraid to say any more, afraid to make the next point. He was worried about the unknowns. He worried that I might indeed find this family of mine, of Samuel Lowe's—find whoever might still exist. If I did manage to track them down, there was the unknown response: how they might react to me. There was a possibility, in fact a likelihood, that my Chinese relatives would not share my enthusiasm for discovering long-lost relatives—long-lost African American relatives. They might see an African American woman and reject me. Until this moment, that thought had never occurred to me. I sat in our hot tub—with my ginger-brown skin, my proudly worn Afro—and, as my glasses fogged from the steam, wondered first why he was saying this to me and, second, shockingly, why I hadn't thought of it on my own.

I paused and inhaled slowly. "I expect that because I am their family, and they are my family, we will be family. That is all I expect."

He nodded gently, with a soft look in his eyes. And at that

moment, I realized his question came from his love, from his wanting to protect me, and from his experience as a man who had grown up in a racist, racially divided United States. I realized he was trying to prepare me for a possible disappointment, to soften a possible blow, to get me to bring down my intensity a few degrees.

He seemed to understand that time would provide answers to the questions that hovered in my mind. He was saying to me: You are passionate about this. You really want it. But in the end, it might not be what you want. It might not turn out to be what you are dreaming. And I knew he was trying to help me. To watch my back. To be the husband I have relied on for decades.

DREAMS OF
MY GRANDFATHER

WE GREW UP IN AN ATMOSPHERE CHARGED WITH THE SENSE THAT SOME-thing was desperately wrong. We lived with a fundamen-tal contradiction: Family is the most important thing in the world, my mother would insist. But we did not have much of a family. I always wondered: if it is the most important thing, why aren't there more of us?

When I was a little girl, I would talk to my grandfather, even though I did not know much about him. I would ask him, "Where are you? Why did you leave?"

And I promised myself that I would find him.

I have been dreaming of my grandfather, of my mother, of the life that they both could have led had they known about each other—and of the life that I would have had if he had not disappeared when my mother was three years old.

Or was my mother the one who had disappeared from his life?

I practice saying the word "Grandpa" out loud. Here I am, sixty years old, and I am playing with a word that I have never used except to refer to a third party. When my grandson Idris plays with his grandpa—my husband, Roosevelt—the word seems simple, easy, and right. I love watching them play together and am struck by how beautifully they take each other for granted, in the nicest possible way. Of course Idris has a grandpa. Of course Roosevelt had a grandpa. That is a fact of life.

But it's not a fact of my life. When I say "Grandpa" aloud, as if this man about whom I have been dreaming might answer, the word sounds awkward and strange. I feel as if I were speaking another language, or wearing clothes that belong to someone else, or calling someone else's grandfather my grandpa.

A powerful undercurrent draws my thoughts to images of my mother, the family that she never knew, and the family that she created. We had only each other back then in Harlem. The four of us: my mother, Nell Vera Lowe Williams; Elrick; Howard; and me. When I think of our family, I don't usually include our daddy, but sometimes I do.

Growing up, we had Cousin George, Aunt Rose, and Uncle Hugh. They were our relatives. That was it. No one else, not for years and years. We did hear of some people in Jamaica—half brothers and half sisters, stepbrothers and stepsisters of my father's, a half sister of my mother's—but they were so remote as to seem unreal.

I would go to the apartment of a friend's grandmother, in the neighborhood, and it would smell as a grandma's house should smell: warm fragrances of pies and chicken dinners and maybe some sweetly scented grandma soap. But what did my grandma's house smell like? Plantains, curry goat, pimiento, ginger?

And what did my grandpa's house smell like? Salt-baked chicken? Preserved mustard greens? Stuffed tofu? Soy sauce?

What house?

What grandpa?

My mother looked Chinese and my brothers and I look African American. We were already separate and different; and when you are different, and don't have a family around you, you feel out of step. As we walked through the streets of Harlem—on Amsterdam Avenue or on 164th Street—men would admire my tall, elegant, exotic-looking mother, and the three of us would stand close by, glaring at them for having "that look" in their eyes when they saw her. We knew we were not like our neighbors, not like the kids in our classes at Saint Rose of Lima Roman Catholic School, not like the folks sitting on their front stoops in the summer heat. That knowledge gave us strength, solidarity, and a particular pride. But it did not fill the yearning for something more. It did not make us feel as if we hadn't, somehow, been amputated.

PART ONE

MOTHERS AND DAUGHTERS

... you see love liberates. it doesn't bind, love says i love you. i love you if you're in China, i love you if you're across town, i love you if you're in Harlem, i love you. i would like to be near you, i would like to have your arms around me i would like to have your voice in my ear but that's not possible now ...

—MAYA ANGELOU, "LOVE LIBERATES"

BEIJING, 2008

I WAS ALREADY AN EXECUTIVE VICE PRESIDENT AT NBC UNIVERSAL IN 2008, when the Olympics were held in China. I had traveled to the games with some of the other executives. We were removed from the day-to-day operations of actually getting the events on the air, but we were seriously involved in an informal diplomatic mission from the United States to China.

As our plane circled over Beijing that August morning, preparing for landing, I glanced out the window and saw beneath me the wide landscape of China, a country that seemed not even to have a horizon; it extended so far beyond any dimensions that I had yet seen. To me, this was both a fact and a metaphor. "Somewhere in the huge continent of Asia, in this vast country of China," I thought, responding to both the physical and the emotional dimensions of the place, "I have family."

Given my appearance and my history, I should have felt this way when I went to Africa. I am, after all, to the world an *African* American. I have been to Africa more than half a dozen times; in fact, I started my travels there, in South Africa. When we landed in Johannesburg in 1998, I was weak in the knees as I got off the airplane. Like so many African Americans, I had a sense of finally arriving; and this continent of history and myth, liberation and domination, purity and contamination, evoked a complicated stew of emotions: sadness and joy, peace and anxiety, connection and alienation. I looked into African faces, hoping to see my own reflection.

But I didn't.

I was truly happy to be there, but I didn't feel as if I had found my spot in this world. It was different when we went to Ghana several years later. I learned only afterward that many Jamaican immigrants were descended from the Ashanti people of present-day Ghana. When I was there, people's faces seemed familiar to me; indeed, I thought these people *look* like me. So did people in Jamaica, and some folks in Harlem. Still, the similarity of appearance did not give me a sense that I was connecting with my ancestors. I was simply connecting with, well, people who resembled me. I assumed, when I arrived in China, that the Chinese faces I saw on the streets of Beijing would have no relation to me at all.

One day, after the opening ceremonies were over, I took a walk in downtown Beijing. We were staying at the St. Regis Hotel in the heart of the city. I thought that I would walk past the embassies and office buildings there, and perhaps visit the Silk Market a few blocks away. The crowds seemed phenomenal, even to a New Yorker like me. As I walked, I absorbed the energy and the intensity of the place; my reportorial instincts snapped into action as I looked at the details of the people, the cars, the bicycles, the storefronts.

And then I saw a face that stopped me cold. I turned my head and watched the woman stride past me and disappear into the crowd. She was tall and graceful, walking with a kind of determination and detachment that I knew very well. Her walk made it clear that she was a woman to be reckoned with—whether in the streets of Beijing or on Amsterdam Avenue in Harlem—but no one should dare to try. It was her face that had captured me in a split second. There, in the mass of people on the streets of Beijing, I saw my mother's face.

My mother had been dead for two years, as had my father. So the small band of Williamses, which had become more numerous with marriages and the birth of children and grandchildren, still seemed nonetheless incomplete. But at that moment in the streets of Beijing, seeing my mother's face, I realized that in death she might lead me as she never could during her life. Perhaps here in China, I would unravel the mystery of our identity that resided in our DNA, in our mother's essential loneliness, in the contradictions and achievements that defined us.

THE OUTSIDE CHILD

T STARTS WITH HOW I LOOK: BROWN SKIN, UNMISTAKABLY "BLACK" NOSE, nappy hair that I once wrestled into submission and have now, for many years, liberated to curly chaos.

My older brothers and I were born and raised on Amsterdam Avenue between 163rd and 164th Streets, a neighborhood that sent very few kids to college but many to jail. Years after we had grown up and left, our block—with its boarded-up windows, abandoned storefronts, and glassy-eyed men just hanging out—enjoyed the grim distinction of having the highest crime rate in New York City.

But we always knew we were different, set apart, not like the other kids.

Most of the kids we knew came from sprawling families with a tapestry of cousins and aunts and uncles. The really lucky ones had a grandfather or a grandmother, maybe even two. These families would feast on food, like ham hocks or collard greens or pigs' feet, that never graced our table. We had rice at every meal—a starch I have loathed for most of my life and can now stand in only small amounts—and exotic vegetables no one else ate, like bok choy. During the 1950s and early 1960s, when we were growing up, most other families were intact, with a father and a mother in the apartment.

Only four of us lived in our ground-floor apartment on Amsterdam Avenue—my older brothers Elrick and Howard, my mother, and I. My father, Elrick Sr., lived in Springfield Gardens, Queens, and was alternately engaged with our family and alienated. We were a tiny band of Williamses; I remember always feeling as if there were simply not enough of us.

Most other kids had families who would escape the city summers by getting on a bus and going to Georgia or North Carolina for big reunions with more cousins, more aunts and uncles. Perhaps the crown jewel, a grandfather, would command the head of the table.

"When are we going Downsouth?" I would ask my mother, as we sat on the summertime stoop of our apartment building. I was five years old and bored out of my mind, because the entire playmate population of Harlem seemed to have gone to this wonderful place called "Downsouth."

"We are not from down South," she would reply.

"Ma, *everybody* is from Downsouth," I would insist.

"We're not," she would say.

"Well, where are we from?" I would ask, already sick of the conversation.

"We are from Jamaica," she would say. "We're from an island."

"Well, when are we going there?"

"We aren't going there."

And that was where it stopped. Was there only a one-way ticket away from Jamaica? No return? Or did she mean to imply that there were no means of transportation available? I tried another approach.

"How did *you* get here, Mommy?"

"I came here on a plane."

"Well, why can't we . . ."

"Because we don't have the money," she would say. Case closed.

But while money was important—clearly, you couldn't do much without money—she did not bother to point out that even in Jamaica there was not much family with whom to reunite.

I grew up believing that both my grandfathers were dead, and that my grandmothers—though I knew they existed—were not well regarded by my mother. To ask my mother about her people was to hit a brick wall: the conversation, like the one about Downsouth, went nowhere. But this one was pervaded by a sense of loss that I could feel but could not fathom and would certainly never be able to discuss, let alone probe. Her mother was in Jamaica; we knew that. But where was her father?

"He went back to China and died," she would say.

I have spent my life—my hardworking life—with a persistent and painful sense of loss. My yearning for clarity, my longing to feel somehow complete, was never satisfied by my prickly, demanding, beautiful, fair-skinned, "half Chiney" mother. How could it be, when she walked through her life, and the lives of her husband and children, as a woman of chronic impatience, enveloping depression, and ambient loss?

Once, when I was a reporter for the *Syracuse Herald Journal* in upstate New York, my daughter Imani and I were sitting at home. She was three years old and was gently stroking my hair, which was washed, combed out, and smooth. "When will my hair be like yours, Mommy?" she asked in a soft and thoughtful voice. Her halo of curls glistened in the light and I was startled by her question. "Your hair *is* like mine," I said.

"No it's not, Mommy," she replied sadly. "I want hair like yours."

I never remember saying such words to my mother, although the difference in our hair was a defining experience for me; despite our attachment, it encapsulated our alienation from each other. My mother always looked more like her Chinese father than like her Black Jamaican mother. She was unusually tall, very slender and graceful, with jet-black hair that would have tumbled down to her waist, had she removed the big hairpins that kept her French twist tightly anchored. She was baffled by my hair, unable to cope with the tight curls and coarse texture, and would outsource its care to an upstairs neighbor, a sweet woman from Downsouth—North Carolina, to be exact.

I always resented my mother for that. I remember seeing other little girls in our neighborhood sitting on the floor, or on low wooden stools, or on thick Manhattan telephone books, leaning against their mothers' shins, while the mother would carefully comb through their curls. I could feel the rift between my mother and me widen as her inability to cope with my hair seemed a refusal to cope with me. Her not "knowing" my hair, not sharing those moments with me, was of course a fact of my life. But it was also intensely symbolic; it meant that she did not, could not, appreciate who I was. When I was about four or five, my hair was long enough for my mother to routinely comb, braid, and twist it, but the memory of those very early years, when she didn't, lingered.

My relationship with Imani—and hers with me—had all the ease and tenderness that was lacking between my mother and me. When there was a potential rift with Imani, I would be sure to bridge it. The day after my sweet little girl and I had our conversation, I had my hair cut short in an Afro, and I kept it that way until her hair grew longer than mine. "Now," I said. "Now do you see? I told you that you and I had the same hair." She touched first my hair and then hers, and a radiant smile illuminated her face. For Imani, this made sense. But when my mother saw me, she screamed. "What have you done?" she cried. "Oh, my God! You look like a pickney!"

That was the last time she ever called me a pickney—in Jamaican patois, a pickaninny.

Mothers and daughters. Or more to the point, *my* mother and *my* daughter. I don't know any African American woman who has not had, at least once in her life, some issue with her hair. Its texture, cut, and color; the decision to straighten it or to let it assert with pride its occasionally impudent personality—all this comes with the implication that whatever we do, our hair will provoke some reaction or judgment. But the negative judgments usually came from someone other than your own mother. My mother's reaction to me was a symptom of something fundamentally amiss in our relationship, a basic lack of identification between us. I was always my father's girl, and this fact seemed to inflame my mother and provoke a longing in her. "You don't know what it is like to grow up without the love of a father," she would say to me.

She was right. But when I was young and she said this, I heard only an annoying bleat. Her complaint was repeated too often for me to take it seriously. It was as persistent and as easy to ignore as the hairline crack in the kitchen wall, a predictable rebuke that was just part of her repertoire. But sometimes, when we grow up, we look back on the things we heard too often, and try to hear them as if for the first time.

Nowadays I am hearing her for the first time.

And I find myself trying to imagine my mother when she was a girl.

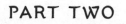

PART TWO

WHERE LOVE BEGINS

The same mouth that courts you
doesn't marry you.

—CARIBBEAN PROVERB

KINGSTON, JAMAICA, 1918

K INGSTON HAD BEEN THE CAPITAL OF JAMAICA SINCE THE END OF THE SEV-
enteenth century, and despite the ravages of storms and earth-
quakes, it prospered. The population soared from 1850 until
1920 as white imperialists, many from Great Britain; other island
folks; Chinese; Indians; and more crowded the small port city. This
influx of humanity also brought mass unemployment and areas of
squalor and crime. One observer wrote, "At the best of times . . .
Kingston bore a foul odor of sweaty human bodies moving about in
the tropical heat, of rotting garbage, the carcasses of dead animals,
and sewage in street drains."

With its wooden houses and utilitarian architecture, Kingston may
have lacked the charm of Spanish colonial towns, but nothing could
beat its natural beauty. To the south is the seventh-largest natural har-
bor in the world, and the Blue Mountains surround the northern part
of the city. Kingston became the most important trading center in the
Western Hemisphere in the eighteenth and nineteenth centuries.

The city was all but destroyed by the earthquake of 1907, and
what the tremors and shocks did not destroy, the ensuing fires oblit-
erated. And yet, for some, the destruction offered an opportunity to
rebuild and beautify the city. By 1917, a new Kingston had risen from

the ashes, sustained in part by the influx of Chinese and other entrepreneurial immigrants making their mark on the city.

My mother was born in Kingston on November 15, 1918, the first
child of a Chinese businessman named Samuel Lowe and a Jamaican
woman named Albertha Beryl Campbell. One of the most important
rituals at birth is to "duppyproof" the infant. There are ghosts and
malevolent spirits everywhere in Jamaica and the culture has come up
with all sorts of ways to either fight or protect themselves against these
invisible but potent presences. A duppy could be the soul of a dead person or some other spiritual being, and it could wreak havoc on the soul,
even the life, of a helpless infant. So parents duppyproofed a new baby
by, say, having one name on the birth certificate but never using it.

My mother's name is Nell Vera Lowe. I found her birth certificate,
but not easily. We looked her up under her mother's surname, as Nell
Vera Campbell. We looked her up as Nell Lowe and as Vera Lowe.
Nothing. And as the daughter of Albertha, Alberta, Bertha, even
Beryl Campbell. Nothing. She was as duppyproofed as one can get.

It might be unnecessary to say here that Samuel Lowe and Albertha Campbell were not married. It's strange to think that both my
parents were out-of-wedlock, illegitimate, or, as they are known in
Jamaica, "outside children." My aunt Ouida, who is one of my father's
stepsisters, once explained, "When you heard about an 'outside child'
you knew that it was not a child that was part of a marriage, or part of
a family that originated from parents who were married." My father
had many adults in his life, perhaps even too many at some point; but
my mother was fairly isolated, starting at a very early age. She shared
little about those times, maybe because she didn't recall, or maybe
because the life she had was not one she wanted to remember.

Her father, Samuel Lowe, eventually became a prosperous shopkeeper in Saint Ann's Bay in Saint Ann Parish. But he began his retail
work by opening his first shop in Mocho, in Clarendon Parish. Mocho
was a place so remote and backward that saying people were "from
Mocho" became shorthand, in Jamaica, for saying they were practi-

cally barbarians. Samuel Lowe's shop was a kind of small general store, called a "Chiney shop." As a storekeeper, he was thoroughly integrated into the community of Mocho. As a young, unmarried man, he naturally developed some romantic relationships with local women and he was involved with at least two around the time of my mother's birth. My mother's mother, Albertha Campbell, was one, and Emma Allison was the other. It is likely that neither woman knew about the other, even though they both bore baby girls at the end of 1918. My mother was born in November and Emma's daughter Adassa was born in December.

I suppose that a man with children from different women would divide his time, although as a very small child, my mother lived with Samuel and Albertha in a small shop he had in Kingston. The family would eat together, and Samuel would play with his little daughter, speaking to her, and teaching her to count, in Hakka, the native language of his Chinese ethnic group; and giving her small treats from his store. This was during a time of rising tension within the local community between the Blacks and the Chinese immigrants. Jamaican nationalism was just beginning, and this early phase included the work of Marcus Garvey—a Jamaican descended from the legendary rebellious slaves known as Maroons. Garvey's worldwide campaign for Black Nationalism began around 1920.

After Nell was born, Albertha continued to manage Samuel's shop in Kingston. Samuel was industrious. He had come to Jamaica to make money and so he also had the shop in Mocho, managed by Emma Allison, the mother of his other daughter, Adassa.

For the uneducated residents of Jamaica's outer "country" parishes, the international movement of Black Nationalism was less important than the personal experience of seeing Chinese folks making money while Black folks stayed poor. Actions taken against the Chinese resulted from a toxic combination of genuine aspirations and acute class envy. Various free-floating resentments were galvanized by incidents large and small. Men regularly harassed women, like

Albertha, who audaciously had a baby with a "Chineyman." There were frightening incidents when Chinese people were blamed for crimes they had never committed, and beaten for infractions that had never even occurred. Sometimes shops owned by the Chinese were burned or a Chinese businessman would be murdered.

Albertha endured the danger and the slights, because Samuel gave her and her daughter a life she would not have otherwise known. She managed and ran his shop and had the security of knowing that while she was not "the Mrs." she was "the missus." But Samuel seems to have told her in 1920 that his relatives in China were sending him a Chinese bride from the prosperous Ho family. Swee Yin Ho was twelve years his junior and their families had "engaged" them sight unseen. As a traditionalist, Samuel assumed that his Chinese bride would raise his half-Jamaican children alongside the children that he and she would have together.

Albertha felt betrayed. She was already enduring verbal abuse and sideways glances from her bumpkin countrymen and -women, and Samuel's decision to move on was a soul-deep insult. Now she was ridiculed for "laying down" with a Chineyman. Enraged, she took Nell away, permanently. Mother and daughter traveled into a countryside that was a mystery to Samuel but would become home to Nell. She was deposited there with an unloving grandmother, secreted in the countryside and hidden from her despondent father.

Albertha may have been furious, even heartbroken, but she was nothing if not resilient. Young, attractive, and—most important— unfettered by either a man or a child, she was free. She soon left on a succession of adventures, appearing and disappearing unpredictably. She moved on and up in Kingston, while Nell grew up unloved and unwanted in a world of people whose reactions to her ranged from mere indifference to active hostility. This must have been where Nell's pain began—in the countryside with her grandmother, her father inexplicably inaccessible, her mother a fleeting presence, and no chance for a real education.

My mother was Albertha's first child, but when Nell was about five years old, another little girl appeared. Her name was Hyacinth and she was Albertha's second daughter, a Black Jamaican without a trace of Chinese blood. I never heard anything to suggest that they grew up as a cohesive family unit; and my mother's fierce insistence that my brothers and I remain as close as possible probably emerged from her persistent sense of being alone in the world—something she would not wish on anyone. She was an "outside child" in so many ways. She was a child whose father had disappeared. She was a child who didn't look like anyone else.

Albertha's disappearance meant that Emma Allison now reigned supreme as Samuel Lowe's Jamaican companion. She now had three children—first a son with Samuel who died at age six; and two more children by Samuel, Adassa and a little boy named Gilbert, who was born in April 1920. This part of Samuel's family lived in Mocho, where Emma ran Samuel's shop, located at a bend in Mocho Road. The small neighborhood is still called "Hong Kong" because its little shops and rum bars were once owned by Chinese men.

But this was all before the arrival of Samuel's Chinese bride. Gilbert was still nursing when Swee Yin Ho came to Kingston to marry Samuel. It was not uncommon at the time for children of Chinese extraction born in Jamaica to be sent to China at around ten years of age, to become accustomed to their Chinese identity; they would live in families headed by women—the Chinese wives who had been left behind in Guangdong Province. Samuel might have learned from his experience with Albertha, because he negotiated with Emma Allison for custody of both Adassa and Gilbert. In exchange for his two children, Samuel gave Emma a lovely, sturdy house on the highest hill in Mocho. She may have gained a house, but she lost a part of her soul when Adassa left her love and care to be raised by another woman, a "new" mommy.

Swee Yin told her new husband that she would gladly accept Adassa as her own, but she refused to take the infant boy, Gilbert.

Adassa, like Nell, inherited their father's Chinese features and skin coloring; she looked as though both parents were Chinese. Gilbert, however, looked more like Emma, with traces of Chinese features but skin that was positively Black Jamaican; any observer could tell that Swee Yin had not borne this son. Samuel pleaded with Swee Yin but she was unyielding. "Leave the boy to be raised by and to care for his mother," she insisted. And that was that. There may have been another reason for this as well: Swee Yin insisted that *she* be the mother of the oldest son, a position of status and importance in Chinese society.

After his marriage, Samuel left Mocho and Kingston to open a bigger shop, the one in Saint Ann's Bay. This city in the north-central part of the island was part of Jamaica's largest parish, and its location and verdant vegetation make it one of the island's most beautiful areas. (It was the home, incidentally, of the singer Bob Marley.) Samuel's younger brother Sih Chiu, whom he had sent for from China, assisted Samuel in his business.

Samuel had dreams of establishing a business that was to be called Samuel Lowe & Brothers. He and Sih Chiu hatched a plan to bring Gilbert to China; Sih Chiu was planning to return home to China several years hence, and he would bring Gilbert with him. Samuel would have his firstborn son with him come hell or high water. Sih Chiu gave Emma money for Gilbert and looked in on him. But several years later Sih Chiu would die unexpectedly in Jamaica, leaving Gilbert behind and ruining his father's plan to reunite them.

Gilbert and Nell were the only two of Samuel Lowe's eight surviving children who never knew the love and care of their father. They also never knew each other. They were half siblings, living in a small community, but never even knowing that they were kin. In 1927, when Nell was nine and Gilbert was seven, their father and his Chinese wife departed for Hong Kong with nine-year-old Adassa and their three racially Chinese sons, who had been born in Jamaica: Chow Ying, Chow Woo, and Chow Kong.

My mother and Gilbert never saw their father again. Adassa

never saw her mother or her brother Gilbert again. And the pattern of loss in our family began.

Nell's life was hard, demanding, full of endless, arduous, work. She went to school—she said that she managed to get to the second form, which would be the seventh grade in the U.S. system—but then her grandmother pulled her out of school to start earning her keep by cleaning houses and working on a farm. She milked cows—an activity that the three of us who were born and bred New Yorkers considered the stuff of children's books, not real life. But my brother Howard recalls that it was not the hard farmwork that so distressed Nell—my mother never recoiled from hard work, that gene was bred in her. No. What was painful about this particular chore was that to get to the cows, she had to pass a small ragtag group of male farmworkers. She didn't say why, but she was a vulnerable girl without a father, or a brother, or a grandfather, or an uncle in her life to protect her.

When she was twelve, she was raped. This was a story that she shared with me—not my brothers—only to impress on me the dangers of boys, of sex, of being an "alone girl" who could be easily overpowered and damaged forever. Nell may have told her grandmother about the rape, but her grandmother didn't do anything to protect her, or to seek revenge or even simple justice.

Nell continued to work and work hard. She learned how to sew, eventually at the level of a master seamstress. She cut patterns, designed clothes, and made a living with needle and thread. But one of her greatest complaints, her real sorrow, was that she never finished her education. My mother was a brilliant woman, but she was poor, an "outside child," and marginalized in all parts of her life.

Nell must have felt pulled by the same strong undertow that I have felt in my search. She needed to know. She needed to find her father. In 1933, when she was almost sixteen years old, she sought out and found her father's shop on Main Street in Saint Ann's Bay. She must have prepared carefully and must have worn something special—since she was a gifted seamstress, she was able to make even a remnant look elegant. She may have rehearsed her words, imagining a sweet reunion with her father, the businessman Samuel Lowe. She walked into the shop, where the smells of spices, dry goods, clothing, and cooking oil mingled to create a special perfume remembered from long, long ago, from her infancy and early childhood. Two Chinese men greeted her—her uncles Hin Chiu and Sih Chiu.

When they saw her, were they startled by her appearance? Did they see traces of her father in her angular face, her wide-set eyes? Perhaps in her bearing, her height, and her grace? Then she spoke and asked if her father was there; she was looking for him. They must have been astonished to finally see this girl, about whom their brother had spoken for over a dozen years, appear on that lazy hot afternoon. Shy but dignified, she had a presence that she probably wasn't even aware of.

Where is my father? Is he here?

Had she come six months before, the answer would have been yes. But she was too late. *Your father has been searching for you for such a long time,* they must have said. *He loved you and missed you. He wanted you to be with him.*

But was he there?

The answer was no. He had just returned to China and was not going to come back. His time in Jamaica, her uncles told her, was over. They gave her a pair of pearl earrings, which they said were from her father. He would have wanted her to have them, they told her. And when my mother turned and left the shop that day, she did not think that she would go to China to find this man. China was literally ten thousand miles away. She must have felt more alone than she ever had in her life.

The chapter of her life that included a father was closed forever.

USURY AND OBEAH

JOHN HENRY WILLIAMS, KNOWN THROUGHOUT KINGSTON AS "JACK," WAS accustomed to living by his wits and getting what he wanted: cars, houses, fine clothes, and lots of money. A wealthy man, he made his living as a usurer, whose role in an impoverished community, where banks were inaccessible to most people, ranged from benefactor to exploiter. My father's cousin, Charlie Meade, said, "I remember they called him the Money Man. Mr. Williams made Cousin Elrick the rich cousin, the son of a big man."

Several decades later, in the 1930s, one of Jamaica's first popular political leaders, Alexander Bustamante, began as a usurer or moneylender. He could pass for white and he spoke like a "man of the people," but he was really middle class. Like many Jamaicans, he had migrated abroad seeking work, in a circuit that included Cuba, Panama, Costa Rica, and often the United States, and this experience gave him advantages when he returned home. So Jack Williams's profession had a certain prestige.

Jack Williams knew how to play the odds and come out on top. He may have been a rather successful usurer, wheeling and dealing from his office on Love Lane, but the side business that helped to capitalize the moneylending profited from other people's superstition and his own ingenuity. He practiced obeah, a special version of what African Americans call "roots" that existed throughout the Caribbean and West Indies. This was initially an indigenous form of healing and spirituality, which eventually became a way of taunting the colonial powers.

Jack figured out how to play on old superstitions and turn a profit. Someone in need would come to him seeking help to resolve a problem, or fix something that was damaged, or perhaps cure an illness. He would listen to his potential client, tease out the need, and dispense some sage advice. But the resolution of the problem depended on a final step, a way for the aggrieved, the physically or mentally suffering client, to appease the spirits by making an offering, preferably in cold hard cash. Jack would advise the client to put a few shillings—five and ten were nice round numbers—in a small purse during a full moon. After doing this, Jack said, you must walk fifty paces from a particular pimiento tree and at a precise time of day—maybe fifteen minutes past midnight—and the purse must be buried at a particular spot near the tree. After the money was buried, a moment of prayer was required to calm the spirits and complete the offering. The next morning, his spirit ready to be appeased, Jack would go and dig up the money. It became the capital he lent at usurious rates to others in the community.

Jack Williams was my Jamaican grandfather.

MY DADDY ELRICK

ON AUGUST 25, 1917, A SWELTERING, SATURDAY IN KINGSTON, JAMAICA, A fourteen-year-old named Sarah Lloyd went into labor. Sarah was a naive country girl from Mocho. Perhaps she had been one of the children who went into Samuel Lowe's Chiney shop; if so, she may have caught a glimpse of the proprietor and may even have received a small treat from him. It's intriguing to imagine a chance encounter between these two strangers who, decades later, would have three grandchildren in common.

Sarah had been summoned sixty kilometers, thirty-seven miles, east to the big city by her older cousin May in the fall of 1916. It wasn't unusual for city dwellers to send for country cousins, giving them a chance to escape the red clay farms and experience a more refined environment. So Sarah left her mother and three siblings, and arrived in a city that was teeming with activity, with vitality, and with more people than she had ever seen in her life. Sarah could sew and do other chores for her cousin; and May had, in Sarah, both a kinswoman and a servant. May had not bargained for the other role that Sarah would play in her life.

May had married well: her husband was Jack Williams, and they lived in relative splendor. But in a community overrun with children, they were childless, and childlessness was a kind of poverty—every bit as conspicuous as any other kind. It must, then, have been a source of conflict and pain to May and Jack. Shortly after Sarah arrived, Jack seduced her—actually, he committed statutory rape—and she became pregnant. Had Jack been so desperate for an heir? Had Sarah Lloyd been too alluring for him to resist? Was he so frustrated with May that any other woman could have satisfied his needs? Or was he simply a man who played by his own rules? Whatever the explanation, May had not expected this outcome; she had simply expected her young cousin to help around the house.

Even before this humiliation May was impatient and nervous, and something snapped when she learned of Sarah's pregnancy. She sent the girl off to live with other relatives in the city, removing temptation from Jack and, for herself, the constant reminder of his betrayal. But she stayed involved. On a late-summer Saturday, at 9 Hannah Street in Kingston's Hannah Town—a part of the city where, today, police stations are set on fire and gangs dominate— Sarah went into labor and the midwife delivered the baby; May was there as well.

For all babies, including the illegitimate offspring of scandalous liaisons, certain rituals were likely to be observed, in deference not just to the physical surroundings but also to the spirit world, which is such a solid presence in Jamaica that one can virtually lean on it. An open Bible, for example, had to be in the room. Castor oil would be rubbed on Sarah's belly. And when the baby boy emerged, healthy and screaming, his umbilical cord would be cut with a knife that would eventually be buried with the cord itself. The little boy was cleaned and then anointed with blue dye, and his navel was rubbed with nutmeg. Many a midwife smoked a clay pipe, and after the birth, she would gently blow smoke into the baby's eyes. And babies were also duppyproofed.

In my father's case, the duppy was confounded altogether by

means of two different birthdays. My father, Elrick Mortimer Williams, was born on August 25, 1917; but "Elrick Mortimer Lloyd," his duppyproofed alter ego, according to the registered birth certificate, was born on September 16, 1917.

May waited until she was able to take the baby home and present him to her husband. If she had hoped that this implicit accusation would lead Jack to reject his bastard son, she must have been sorely disappointed. Jack was delighted. It was he who named the boy Elrick Williams, and he insisted that Elrick would be raised as their own child and his heir. All of a sudden, May was a mother, to a son whom she hated as the embodiment of her husband's infidelity. She took out the rage she felt toward her husband on his sensitive son—at least when Jack was not looking. My father would always remember that his life at home had been split between abuse from May and lavish attention from Jack.

Sarah remained an outcast but was often permitted to visit her son. As Elrick grew up, he believed she was his "aunt," a kind, loving, and gentle, though unpredictable, contrast to his terrifying mother. May, however, was the constant presence, a woman who would have no qualms about beating him whenever the thought occurred to her.

He attended two schools in Kingston—Kingston Technical School and Kingston Tutorial College—but Kingston Technical School had the greater influence on him. Founded in 1896, KTS had an illustrious history in the community, providing education from elementary school through high school and focusing on the trades. It was a coeducational institution; girls learned to be teachers or studied domestic sciences or handicrafts, while boys studied engineering, machine shop, welding, or electrical installation.

Jamaica was still dominated by its British landlords; independence (which would come in 1962) was decades away. Students who took commercial subjects could therefore prepare for examinations at the Royal Society of Arts in London, while those who took technical and domestic science courses would be able to participate in the City and Guilds Institute in London, Britain's largest vocational education institution.

Elrick was a brilliant student at KTS, and especially gifted in math. When I was small, I watched as my daddy's friends called out ten to twelve multiple-integer numbers and asked him to multiply, then divide, then extract the square root of the result. In a few moments the correct answer would pop out of his mouth. Amazed, I would wonder, "Who *does* that? How can *anyone* do that?" Later, when we were in elementary school, I would see my oldest brother, Elrick Jr., doing the same thing. I concluded that I was the abnormal one and everyone else was like my father and brother.

The certainties of math must have been wonderfully reassuring to Elrick Sr. when the craziness of his autobiography became increasingly evident. Fifteen is never an easy age, and for my father it was complicated further by two shocking and life-changing revelations. First, he learned that his aunt Sarah was really his mother. When he realized that, the second revelation came: he also understood the unseemly details of his conception.

To realize that his mother was only fourteen when his father seduced her must have come as a terrible shock. He needed only to look at the younger girls in his school to imagine how inappropriate his father's behavior was. And yet he adored his father—and as unsettling as these revelations were, they were also liberating. In 1943, when Jack died, he was able to escape from May and no longer had to defer to her with the respect that a son was required to show his mother. He also decided to get to know his damaged and equally unreliable mother, Sarah.

THIS IS WHERE
SARAH CAME FROM

MY UNCLE HARRY IS ACTUALLY NAMED ANTHONY HARRISON, AND HE IS MY father's much younger half brother. Uncle Harry was born sixteen years after my father and they shared a mother, but not a father. When Harry was little, he couldn't manage to say the sounds and syllables of "Elrick," my father's name. And the formalities of the time required him to say "Brother Elrick," which was even more difficult. But children are good at collapsing syllables and creating their own linguistic logic, and the names they devise in infancy often last into old age. Harry ended up calling my father "Baba."

Curiously, in many African languages, and in Hakka, "baba" also means father. So Uncle Harry referred to his older brother as father. In many ways the diminutive was also emotionally appropriate; Elrick acted like a father to Harry, disciplining him, cajoling him, trying to make sure that he behaved. This naturally created lifelong tension between the two, a back-and-forth of exasperation, irritation, and affection.

One day, Uncle Harry and I were sitting together at the kitchen table in his home in Fort Lauderdale, Florida, a bottle of Jamaican rum—151-proof J. Wray & Nephew—set before us. As my father had on many occasions, my uncle warned me against drinking it straight—no ice, no chaser.

"Paula," he admonished me, "it's too strong to drink like that. How you and your brothers do that I cannot understand."

"Uncle Harry, don't worry," I said, "I'm good." I didn't want to discuss my choice of a beverage; I wanted to talk to him about his mother, my grandmother, Sarah Lloyd. Uncle Harry paused. "You want me to do that now?" he asked, as if this demand would exact a great deal of time and energy from both of us.

"Sure," I responded. "Why not?"

"OK, then. OK."

And Uncle Harry began.

"My family name is Lloyd, OK? There was a gentleman from Scotland by the name of Lloyd and he came to Jamaica. He had two children by an African woman. One was Fannie Lloyd, a girl, who lived to be a hundred and three, and she had only one daughter.

"He also had a boy with that same woman, and they named him Cefus Lloyd. Cefus met Keturah Brown in Mocho and they had four children—three girls and one boy. The eldest girl was Icilda (Icy) Lloyd, who got married to a man named Barnes. Then the next child was Sarah Lloyd, who was my mother, your grandmother. The next was Peter Lloyd, the only boy in the family. Then the last was Catherine Lloyd who was my aunt, the baby." The baby, Catherine, was called Dada by the entire family.

Keturah—Kettie—was a very kind woman. It was said that she tied her money in a cloth and she treated her grandchildren to a penny. She earned a few shillings by sewing and embroidering, yet she was afflicted with glaucoma and poor vision. Kettie wore dark glasses always—even at night—and her eye disease persists to this day in my family: in my father, my uncle, and my brother.

Cefus and Keturah had their four children in the early twentieth century. They lived in the country town of Mocho, Clarendon, where Grandfather Samuel Lowe started his first shop. Did Samuel's shop count the Lloyd family among its customers? Could my grand-mother Sarah on my father's side have bought goods and bits from "missah" Lowe, my grandfather on my mother's side? Could their son and daughter have married years later in Harlem?

Cefus, a prosperous man of Scottish and African descent who looked almost white, had quite a nice piece of property, according to my cousin Charlie Meade. "I didn't know my grandfather very well, but he lived not too far from us in Mocho. He had a sister who was very, very fair-skinned, like him. She was an evil woman; I used to hear people call her the Old Bitch." Cefus was in love with Keturah, who was sometimes known as Sister Kettie. But his sister Fannie hated dark-skinned people. Charlie recalls that Fannie was kind to her fair-skinned nieces and nephews, like his sister Lunnie and my uncle Harry, but he says, "She hated me because I have dark skin like my grandmother, Sister Kettie." Fannie was the elder sibling who forbade her Cefus to cement his union with the tall, dark-skinned, "ramrod-straight" Keturah. Charlie doesn't think Cefus ever married or had another woman or other children, and he is fairly certain that his grandmother Sister Kettie had no rivals for Cefus's affection.

Sarah Lloyd knew early on that her father lived in the same town, although she'd never seen him, and she felt that she and her brother and sisters were entitled to have a father.

One day, Sarah and her brother and sisters hatched a plan to meet their father by going to his home. They walked along the dirt road until they arrived at his house. "Good day, sir. Howdy do?" said Sarah.

From his veranda, Cefus replied kindly, "Howdy do, miss."

"What a lovely property," Sarah said sweetly, as Icy, Catherine, and Peter looked on. "The master of this home and his family must truly love it, for it is so beautiful." Cefus was pleased by the compli-ments from these well-mannered young people.

Cefus said, "It's my home and we do love it."

"So do we," Sarah told him, and then she brandished her weapon. "And as your children, we should share in its pleasures." From then on, Cefus was known to his children, but the damage was done; a close relationship was not to be. Charlie recalls seeing a man sitting on his veranda one day as he returned home from school. He asked his mother—Catherine, or Dada—who the man was. Aunt Catherine told him, "Oh, he's your grandfather and he came to meet you." Cefus had at least a dozen "grands" by that time and decided that he wanted to know his offspring and theirs.

Uncle Harry said his mother told him this story many times throughout his life. That day, Cefus's outside children had confronted their father, a father who, because of misplaced loyalty to his sister Fannie, didn't recognize his own offspring. Keturah was abandoned, left to raise her three daughters and her son in a fatherless home. Small wonder that she had felt a need to send her daughter Sarah off to live with her prosperous cousin May in the big city.

Uncle Harry leaned back in his chair. There was something about telling this story that both energized and exhausted him. Here he was, a distinguished man in his seventies, still trying to make sense of the difficult mother who bore him.

ELRICK GROWS UP

WHEN ELRICK TURNED TO SARAH, AS HIS MOTHER, HE FOUND HER NOW IN love with a kind and charming man named Marques Harrison, who was quite prolific, it seems, when it came to fathering children. He had three children with his legal wife, Juliet Anderson Harrison—Sarah's rival. Marques also had outside children, for a total of eleven sons and daughters; and when Elrick was sixteen, Sarah became pregnant with Marques's twelfth child. Suddenly, my father had a web of stepbrothers and stepsisters. In the Jamaican tradition, these family ties were maintained for generations. Actually, however, my father was disappointed in his mother's choice of a partner and longed for a better life for her, so he asked his father for permission to give Sarah his weekly allowance. Jack was so impressed with his son's generosity and sense of responsibility that he not only granted permission but increased Elrick's allowance.

When the baby was born, Sarah suggested that Elrick name his new brother, and my father was delighted. The boy would be called Horatio Anthony Harrison, after Horatio Nelson, the British vice admiral who was a leader during the eighteenth-century Napoleonic wars. Elrick, a good British subject, saw Viscount Nelson as a hero and thought that his brother deserved a noble name. Thus my father gained a brother, albeit one who was sixteen years his junior. The difference in their ages, combined with differences in their personalities, created tension, resentment, protectiveness, and annoyance—dynamics that would endure for the rest of their lives. As for Uncle Harry, he legally dispensed with the name Horatio as soon as he could, changing his name to Anthony. And my father never forgave him.

In 1942, Juliet Harrison died, leaving her daughter and two sons—Ouida, seven; Paul, five; and Maxime, three—motherless. Marques married Sarah, and she became their reluctant stepmother. It must have been a rather chaotic existence because Sarah also had her own six-year-old daughter, Carmen, known as Cutie; and the son she bore for Marques, Harry, who was nine. Five children under the age of ten would be taxing for any woman, and Sarah struggled to manage the load. And then there was Elrick, who was a grown man of twenty-five.

Being Jamaican was exciting and difficult during the 1930s. My father and his family kept their distance from politics, but it was all around them. Jamaica was not exempt from the consequences of the worldwide Depression, and throughout the Caribbean anticolonial resentment combined with the dire economic circumstances led to strikes and social unrest. Economic catastrophe, as so often happens, sparked overdue social reform. Migration to the United States and elsewhere ended, and many Jamaicans who had been abroad returned home to participate in the creation of a new—or at least a different, unfettered—country.

Worsening economic conditions led to the rise of new political leaders and labor unions and brought to an end a relatively peaceful period—albeit one when Jamaicans had endured the limitations and humiliations of being colonized. The emotional consequences of the economic collapse were also drastic: suicide rates increased by 25 percent.

My father worked at a number of odd jobs in Kingston, and often his father employed him in the less savory but toughening aspects of moneylending. My father became "the enforcer." As he collected debts, he got to know the city and the full range of personalities who inhabited it. When we were kids, he impressed us with the drama and the danger of his occupation.

Our favorite was the story of the Avenging Bus Driver. A bus driver in Kingston had a side business of lending money. On the way to work one day, he collared one of his long-overdue debtors, a man who had kept putting him off and stalling. They walked along and argued until they arrived at the bus station, where the driver gave an ultimatum: "I am going to run my route, and by the time I get back here, you'd better have my money."

So the man drove his route and when he returned his client was still there, empty-handed. With the cool determination of a professional assassin, the bus driver pulled out a machete and cut the debtor's head off. Daddy always said the debtor was so desperate to escape that the headless body took off and ran down the street.

This was the kind of fabulist tale we came to expect in the Jamaican world, where ghosts, spirits, strange goings-on, and strong statements from the "other side" were as much a part of life as plantains and rum. But on a practical level, our daddy learned to get tough about getting what he was owed, and he passed that lesson down to us. My father was a gambler, an expert poker player, and skilled at all sorts of moneymaking schemes. In the neighborhood, his nickname was "Champ," implying all the riches that others lost and he won.

Many years later, when Jack was long dead, my brothers accompanied our father on a trip to Jamaica. He was an impressive-looking man, a handsome, lean, tall, dark brown Jamaican. He had smooth, fine skin and jet-black hair with a wavy, curly whorl—"good" hair, not nappy, the kind of hair that people on Amsterdam Avenue said could "take water." Despite the heat, he was all decked out in his three-piece suit and "stingy brim felt hat," the haberdashery of a man who was secure in his status and would never, but never, put on a pair of what he called "dungarees" and the rest of us call jeans.

Dungarees were the clothes of the poor, and to my father, they were demeaning.

As they walked down the street, my father suddenly began to chuckle. "See that guy down near the end of the block?" he asked. "Watch. When he sees me he is going to cross the street." Sure enough, not ten paces later, the man's eyes gradually grew wider as he approached the men in my family and then crossed the street, watching our father nervously.

My brothers burst out laughing. "Daddy, what is up with that?" Howard asked.

"He still owes *my father* money and he thinks I'm looking for him," Daddy explained. This was more than thirty years after his father, the famous usurer, had died.

My Jamaican grandfather died when my father was a young man, leaving his son all of his properties, two cars, money, and the attention of many young ladies. May became less relevant in Elrick's life, and he protected Sarah as if she were his child, and not the other way around.

In the late 1930s and early 1940s he drove around town as if he owned the place. He made some money by picking up his father's business, and by finding some extra work of his own. He took his younger brother Horatio—who had become Harry to everyone in the family—under his wing. Despite the Great Depression, despite the hardships that were peculiar to my father's crazy youth, and despite the fundamental lack of opportunity for a man like him, he had a pretty good life. He had some money, and he also was liberated by knowing the truth of his origin. Nothing tied him to May, now a widow. He was free from any filial obligation to the woman who was not his mother.

Elrick's generosity and her new husband notwithstanding, Sarah needed a job and found one as a gifted seamstress at Kingston's prestigious Alpha School for Girls. The school was established in 1880 when a wealthy African-French-Portuguese-Jamaican woman named Mary

Ann Justina Ripoll purchased forty-three acres in Kingston. There was one building on all that acreage, and it was known as the Alpha Cottage. In that building she first housed a single orphan, but also a vision of something that she could bring to the community.

About ten years later, the Sisters of Mercy made Alpha Cottage their home and Mary Ann Justina Ripoll became Sister Mary Peter Claver. The nuns initially brought in orphans from every parish on the island to shelter them and provide them with a home. Then, as Catholic nuns all over the world have done for generations, they created a school to educate these orphans. Eventually it became two schools—the Alpha School for Boys and the Alpha School for Girls—and not only orphans but also children from many Kingston neighborhoods put on the uniform and learned academic subjects, self-discipline, and religion. There was also a private preparatory academy that became an elite institution where the cream of Jamaican society—both Blacks and Chinese—would send their daughters for a proper education. For Sarah, getting a job there was a godsend.

In my grandmother's biography, this would probably not have been much more than just a job, except for one friendship she developed there, with another young seamstress: a beautiful, light-skinned, "half-Chiney" girl named Nell Vera Lowe.

TWO OUTSIDE
CHILDREN MEET

IT MUST HAVE BEEN AROUND 1939 WHEN SARAH, THE HEAD SEAMSTRESS AT THE Alpha School for Girls, took a second look at the twenty-one-year-old half-Chinese woman who started work there, and began to scheme. She learned that the girl's name was Nell Vera Lowe. She learned that Nell's father was not around, and her mother was not much of a mother. She could see that the girl was beautiful and light-skinned, but poor and lacking any of the important social connections that could lead to a good, prosperous marriage—a marriage that would raise her up from her difficult circumstances.

Sarah thought about her son Elrick. He was such a handsome young man, twenty-two years old. Maybe he was running a little wild with the kind of women who would never do him much good; still, he was a good boy. Sarah realized that her son had two major liabilities that would make it difficult for him to find a suitable wife. The first was that he was very dark-skinned. The idiotic prejudice—I suppose all prejudices are idiotic—that has so troubled the lives of Black folks here in the United States did not spare our Caribbean brothers and sisters. Being very dark-skinned put someone on a lower social level than being lighter-skinned. Sarah was very canny about these various shades, the spectrum of desirability.

The second problem was even more vexing. Any woman from a decent family would want to marry a man from a decent family. There was no way to sugarcoat Elrick's parentage. For a child to be born out of wedlock, to be an outside child, was unfortunate, but it could be managed, especially in a country where there were many such children. But Jack Williams was a well-known man, and what had happened between him and Sarah breached the bounds of social acceptability. Elrick carried the taint of that relationship as his own mark of Cain, and there was not much to be done about it.

Not much, except to find him a woman, an attractive light-skinned woman who didn't have anything to lose, and who didn't have a family that would dismiss him as unworthy even before laying eyes on him.

Sarah introduced Elrick to her young, half-Chinese fellow seam-stress, Nell.

Somehow Nell and Elrick found something in each other that res-onated. Perhaps each appreciated the essential loneliness of the other. Or perhaps they shared a sense of being marginal in the world. Maybe my mother felt embraced by my father's sprawling, multilayered fam-ily. He had a half brother and a half sister, and all those stepbrothers and stepsisters who were also a part of his life. Sarah's satisfaction with her own matchmaking must have been expressed by the way she welcomed Nell into their lives. For Nell, this was as close as she had come to a loving family since she was three years old.

The two of them vowed never to have children out of wedlock. They would never expose a child to the kind of life that they both knew painfully well. They had an understanding, the two of them, and eventually they began to live together. Nell sewed and put away money. Elrick worked at his various jobs, for by now he had acquired marketable engineering skills and became a tool and die maker. He was living on his income and his inheritance and he enjoyed having the ravishing mixed-race beauty on his arm.

But Nell was not going to stay in Jamaica. She would not let the island be her prison. The obvious land of opportunity for everyone was the United States, and some of her mother's relatives were willing to sponsor her. Her two cousins were a generation older than she, responsible and established and the perfect destination for her after she arrived. Her uncle Hugh and his sister Aunt Rose Holness had made their way to the land of prosperity and settled on 126th Street in Harlem.

The Depression made immigration policy especially tight, and Nell's special status as a half-Chinese woman was initially a liability. From 1882 until 1943, the U.S. government reacted to the influx of Chinese immigrants seeking their fortune by imposing severe limitations on immigration from China and of people of Chinese origin from other countries to the United States. In 1882 the government passed an Act to Prohibit the Coming of Chinese Persons into the United States, which permitted some restricted access for Chinese workers but also imposed even greater restrictions on anyone else of Chinese origin.

When Elrick and Nell were getting to know each other and then living together, the United States entered World War II. It must have seemed a distant conflict for the people who lived in Kingston. Still, some Jamaican men served in the U.S. Army; some harbors in Jamaica were used for the war effort; and one consequence of the war in particular affected Nell and Elrick. The United States and China were allied against the Japanese. On December 13, 1943, Roosevelt signed an Act to Repeal the Chinese Exclusion Acts, to Establish Quotas, and for Other Purposes. This meant that a quota was set for new Chinese immigrants to the United States. Nell leaped at the chance.

My mother's mother, Albertha Beryl Campbell, still lived in Kingston, along with her youngest daughter, Myrtle Hyacinth Nethersole. Aunt Hyacinth was a beauty and as dark and cheery as Nell was fair and somber. Nell was almost five years older than Hyacinth, who was born in August 1923. The two sisters couldn't have been more different in appearance—understandably, since their fathers were quite literally worlds apart. Nell had her father's Chinese features and Hyacinth was as dark and glossy as her father, a Mr. Nethersole. In Jamaica, a man would typically stay in a woman's bed as long as she made him happy. When he got bored, or she got pushy, or another woman appealed to him more, then he would simply move on and end up in yet another woman's bed. Sometimes he found what he was looking for and stayed. Otherwise, he kept on wandering. Finally, as time went by, he just got too old to care.

Outside children were plentiful in Jamaica. Slavery had destroyed the nuclear family there, as in the United States. In those days, there was little in the way of birth control, and the easy promiscuity of both men and women meant it was not seen as odd for Albertha Beryl Campbell to have two daughters by two men, neither of whom had exchanged marriage vows with her.

I never learned how Aunt Hyacinth was raised but, unlike Nell, she was raised and cared for by her mother, Albertha. The contrast in the two sisters' lives extended far beyond their fathers and the color of their skin. My mother's childhood was marked by verbal and emotional abuse. My father's sister, Aunt Cutie, told me, after my mother died in 2006, that my mother "was so mistreated in that grandmother's house, when her mother and little sister came to visit the country, Miss Nell's grandmother would make her stay in the back of the house so visitors couldn't see her. She was ashamed that her granddaughter was half-Chinese." The family praised and indulged Aunt Hyacinth, partly because she was much younger but mostly because she was pure Black Jamaican and her father was not a Chinese shopkeeper who had disappeared.

Who could blame my mother for wanting to get the hell off an island that held memories of punishing work, cruel relatives, and poverty? Were there any good memories associated with that country? Certainly she never felt affection from nor toward Albertha. She felt responsible for her little sister, Hyacinth, as all elder siblings do.

But the last affection my mother genuinely recalled was when she was living under Samuel Lowe's care in his home at the back of his shop in Kingston.

Father. Mother. Daughter.

Chinese shopkeeper. Jamaican mistress. Bastard half-breed baby.

These are all different lenses through which the people involved are viewed. More important, these are different lenses through which they see themselves.

"Yit. Ngee. Sam. See. Ngh." "One. Two. Three. Four. Five." Samuel Lowe taught little Nell to count in his native Chinese dialect, Hakka. Throughout my childhood, I would sometimes hear my mother speak Chinese phrases. The singsong tones were unintelligible to me and yet they were somehow soothing. Nell taught Elrick Jr. and Howard more: "Luk. Chit. Bat. Ghiu. Sip." "Six. Seven. Eight. Nine. Ten."

At twenty-six, my mother decided that Jamaica had done all it could for her and she was going to make a new life in a new country. She was well on her way to being an old maid at that point, since at her advanced age most traditional Jamaican women were married or at least had children. She chose otherwise. No man had ever done anything for her, she said. She'd always had to do for herself.

Since the United States had eased the immigration policies for Chinese, Nell decided to apply for a visa; it would be her ticket out of Jamaica and the start of a new life. She got relevant information explaining the process: all that was required was a single document to prove who she was and where she had been born. For most folks, coming up with a birth certificate was the easy part of the process. For my mother, it could have been an insurmountable obstacle: she simply didn't have hers and had never even seen it.

She went to Albertha and asked for her birth certificate. And what her mother said left such a scar on my mother's heart that she vowed always to keep the proof of her children's identity, the precious official documents, very close by. Albertha had never registered her eldest daughter's birth. She had never even taken the time to fill out the paperwork with the registrar general to list my mother in Jamaica's records of who was born to whom and when. Nell was not just a bastard child; officially, she did not exist.

After years of searching in vain for my mother's birth certificate, I paid a fee to have the registrar general's office in Jamaica find proof that Nell was the daughter of Samuel Lowe and Albertha Beryl Campbell. I got half of my answer when an e-mail arrived showing Nell Vera Lowe as born "fifteenth November 1918" to Albertha Campbell, "housekeeper." And on the document was the reason why I hadn't found it despite years of searching. Albertha had finally recorded her twenty-six-year-old daughter's birth in *January 1945*.

My mother must have hated her mother for putting her through so much embarrassment. After all Nell's hard work, her mother's disrespect for her lasted into womanhood. This must have been the last straw. Knowing my mother, I doubt that she politely suggested that Albertha leave her residence at 7 Saint John's Road, Johnsontown, Kingston, and travel the few miles to the Saint Andrew Parish government offices to finally record Nell's birth. Knowing my mother, I imagine that she loosed a diatribe at Albertha. Years of neglect, years of disrespect, and years of abuse must have fueled my mother's invective. Nell probably had to stand in the government office with Albertha and prove that she was, in fact, Nell Vera Lowe. Mother named Campbell. Daughter named Lowe. The space on the document for "Name and Surname and Dwelling-Place of Father" is left blank. Why? Why would it not be filled in: "Samuel Lowe, Kingston"?

My mother was twenty-six, almost twenty-seven, and she knew her father's name. Albertha knew her lover's name. But there was no requirement in Jamaica for the name of the father to be recorded, and I can find no legal document that lists both my mother and my grandfather except for my mother's death certificate, which I was in charge of completing. "Her mother was Albertha Beryl Campbell," I told the funeral director. "And her father was Samuel Lowe."

Finally, at the age of twenty-six, almost twenty-seven, Nell Vera Lowe became a recognized, duly registered citizen of Jamaica long enough to hold the document in her hand and fill out the visa application that the U.S. embassy required so she could get away from Jamaica, from Albertha, and, I suspect, from Elrick.

LIES AND MARRIAGE

IN 1945 NELL LEFT ELRICK AND BOARDED A PAN AM FLIGHT TO MIAMI. SHE THEN took the train from Miami to New York City to meet her aunt Rose and her uncle Hugh. For the first and in most important ways the last time in her life, Nell was free. I don't know if there had been violence between Nell and Elrick, or if Nell realized she simply wanted a bigger, better life than the island had to offer. In March 1945, Nell settled in Harlem, a place she would call home, off and on, for the next sixty years.

Elrick was devastated and determined that while she may have abandoned the island she was not going to abandon him. He was determined to share her life, whether in Kingston or in New York. An official visa was out of the question for him—visas weren't being issued to young Jamaican men at the time—so he got a job with a steamship company.

My father told at least two versions of his journey to New York: one for his daughter and one for his sons. He told me that he'd been hired as a merchant marine tool and die maker, a machinist on hand for engine repairs on the voyage north. He told my brothers he was a stowaway. For a few years, I was on the other side from my brothers; we just agreed to disagree. In 2014, I sat with my dad's first cousin,

Charlie Meade, Dada's second-eldest son. "Charlie," I asked, "do you know the story of how my father got to New York and why?"

My cousin gave me one of those eye-twinkling, amused smiles and said, "Of course I know how, Elrick got to New York. We used to laugh together about my description of him as a 'first-class stowaway.'" Charlie told me how, in 1945, my grandfather Jack Williams paid Conrad Sweetland to stash his son aboard the ship and to make sure he was well nourished, well hidden, and well cared for, just like the paying passengers. "Elrick was the only first-class stowaway I ever knew of, but, you see, his father was wealthy and all Elrick could not get was a visa."

Now to understand why Charlie was so familiar with his cousin's travel arrangements, I first had to understand how different his passage out of Jamaica was from my father's. "I stowed away about five years later," my cousin said. "In those days, we wanted to get to the United States, but my best chance—on the days I spent at the docks sizing up the situation on the boats, you know, the police, the cargo, the destinations of the ships—was England." Charlie found that of the ships departing that week, his best option was one bound for England, not the United States. So he wrapped his British passport tightly in plastic and taped it snugly around his leg just below his calf. The plastic was to keep it dry should he have to jump into the sea to escape guards, dogs, or police. Charlie knew that should he reach England without a passport, he would be sent right back to Jamaica. "But with a British passport, as a British subject, they would not send me back. I would be detained for a day or two but then released and I'd be a citizen of Great Britain." Jamaicans were already classified as citizens of the U.K. who lived in a Commonwealth nation.

Charlie said he didn't expect his voyage to be like my dad's. "Elrick was sleeping in a cabin, hidden away, and being served food and drink! What kind of stowaway was that?" He laughed, amused and shaking his head at my father's great fortune. "He came the easy way; he came in comfort, but I came the hard way."

He thought about what he would do once he finally reached London, he said. "But first I had to get past the damn dogs patrolling the docks and on the watch for stowaways." Charlie crouched just off a seawall for hours while the dogs barked and the guards searched for the would-be stowaways. "I was never so scared, because I wanted to get on that ship to England, but if those dogs found me and the other guys hiding with me, we would have been torn to shreds while the guards made an example of us."

It seemed like forever, but finally Charlie and his mates scrambled aboard the ship and hid in the coal rooms. "We must have been in there for a couple of days, man. We were so hungry and thirsty, we couldn't take it anymore. We knew we were out to sea for a long while and they couldn't turn back so we started banging on the walls and pipes and whooping up a lot of noise." They'd decided it was OK to be detected now because the next time their feet touched dry land, it would be the soil of England.

"One of the crewmen who worked the engines heard the noise we were making and came to see what it was," Charlie said. "When that guy opened the locked door and all he saw were figures completely black from the soot and coal, he slammed the door back shut and ran screaming for his life! He thought he was seeing ghosts, spirits." Charlie laughed, tears welling in his eyes. "That guy thought the ship was haunted and started screaming and howling. All he could see was the whites of our eyes and our teeth!"

The "duppies" were freed a few minutes later when the crewman brought reinforcements. "All of them were scared; they can have some very superstitious beliefs on those boats." Superstitions or not, Charlie and the others were rounded up. By the time the entire ship was searched, fourteen stowaways were found; they had secreted themselves on board before the ship left Kingston.

"They locked us near the kitchen and, man, we ate like there was no tomorrow." Charlie laughed again. After they broke the lock to the food storage, Charlie said, they ate meats, vegetables, even desserts. "We ate so good and there was so much food there, they didn't even notice how much all of us who had been starving for three days ate up."

The captain had them searched and found and confiscated their passports. "Then they put us to work, cleaning, scrubbing, doing the work of the hired crewmen—the sailors. Man, we scrubbed decks, cleaned toilets, and they tried to work us from dawn to dusk and past that. I told them I ain't no damn slave. I refused to continue being abused like that and so did the rest of us." Charlie said their "stowaways' strike" lasted one day. "That captain had us locked in the bow of that ship. Now, when you are at the very front of the ship where it comes to a point, you're cutting through the water," he explained, noting that the greatest turbulence would be concentrated right where the ship breaks through the water. "That night, we were tossed around that room, banging on the walls and the floor; just getting beaten up! Man, I never had a night like that before or since." The next morning, the stowaways' work stoppage ended and they resumed their roles as unpaid, reluctant deckhands.

When the ship docked three weeks later, Charlie said he and the others were excited because they had reached England and would soon be heading to London to join fellow countrymen and relatives whose names and addresses they had committed to memory or jotted on a scrap of paper. But not everything went as planned.

"We looked out of those portholes and knew we'd made it." But it wasn't quite what they'd anticipated. "They kept us locked on the ship and we saw the guy who kept the key—the one who locked us up at night, then let us out before dawn—we could see him on the shore. He left the ship and we were locked up!

"Later, they came and took us to jail; to jail!" he said incredulously. They went before a judge who released them late at night to the British countryside. "All the transportation into London was already finished for the night and we had nowhere to go." Disheveled and with a only few pounds each, given to them by the British government, they were directed to a travelers' aid center for down-on-their-luck recent immigrants. Rummaging through the charity's resources, they found used clothes and shoes. "I'd lost one of my shoes when I was climbing on the boat," Charlie said, so he'd gone the three weeks on board without a pair. The next morning, the Jamaican former stowaways made their way to London and were absorbed into the community of fellow Jamaicans in areas such as Brixton.

Charlie's story gave me a clearer picture of what many young Jamaican men had to endure to reach what they saw as the lands of opportunity. "Yes," he said. "My experience was a stowaway's experience—hard." He laughed. "But Elrick's was soft. I never heard of another first-class stowaway before or since." Charlie said he and my dad laughed many times about my father's silver-spoon voyage to America. Elrick was adored by his father, Charlie said. "His father, the rich man, gave him everything he wanted." But it 1945, what Elrick wanted above all else was Nell, and Jack Williams bought his son a first-class illegal trip to New York, where he would again get exactly what he wanted: his half-Chinese beauty who had managed to get away without him.

When the ship docked on Manhattan's West Side, my father knew that my mother was now within reach. He knew he had no authorization to disembark, but leave the ship he did—with a few bills and some coins in his pocket. He caught the subway north to Harlem and exited the train at 110th Street and Lenox Avenue. By the time he completed the fifteen-minute trip, he realized that his demeanor, his unfamiliarity, and his look all made him an easy mark. By the time he climbed the stairs to the street level, he was at a serious disadvantage.

Indeed a pickpocket had already relieved him of the few bills he had. When he reached into his pocket, he felt only the coins. Despondent yet determined, he entered the Parkside Diner—it was still around when I was a kid—and asked the counterman for coffee. Five cents later, over a cuppa joe, the counterman asked if my dad was from Jamaica, because he had noticed the accent. It turned out that the counterman was also Jamaican, and on the spot he offered my dad a job as a dishwasher, starting the next day. Of course, Elrick jumped at it.

Emboldened, my father asked if his newfound friend knew of a Jamaican family named Holness who lived in Harlem. Yes, of course, he knew of them. They owned a couple of brownstones on 126th Street, right off Seventh Avenue. Elrick walked nearly a half mile to find Uncle Hugh and Aunt Rose, knocked on their door, and took the first step into a new life—in Harlem, with Nell Lowe—a life that created us all. He had found his prize.

Yes, he found my mother after only one question to the right man at the counter. How could this not have been meant to be? He had risked everything for her, he said. He had flouted the immigration laws to find her, he said. He wanted to marry her, he said.

He bothered her, she said. That's all. She just gave up arguing with him because he kept bothering her. She should have married that Chinese man who wanted her back in Jamaica, she said. The biggest mistake of her life was getting married and having children, she said.

Years later, when I was myself a married woman, she repeated this refrain: "He bothered me." I looked at her and said, "Ma, you don't really think we ever believed that you married him because he bothered you, do you?" I had changed my lines, departed from our script, disrupted the cadence of our usual back-and-forth. She looked pointedly at me with her lips pursed and expelled her breath in a hiss. This was where the Jamaican in her came out, in making the classic Jamaican woman's expression of disapproval beyond words. Then she walked away.

Elrick married Nell in St. Thomas the Apostle church at 262 West 118th Street in Harlem on September 23, 1945. On the marriage certificate Nell listed an imaginary man, "Daniel James," as her father. I wonder if leaving this space blank might have been a liability. Elrick was playing fast and loose with the truth as well; he listed an incorrect year for his date of birth. Since he was in the United

States illegally, he probably wanted to leave some uncertainty in his official tracks.

Mavis Holness was Nell's maid of honor. She and my mother lived in one of the Harlem brownstones owned by their cousins, Aunt Rose and Uncle Hugh Holness. And my father's address was around the corner from the home of Elrick Jr.'s godfather and his wife, Conrad and Lotte Sweetland. Marriage license 23959 was duly signed and sealed. The couple managed to improvise a family for their wedding at St. Thomas the Apostle Roman Catholic church. Nell had to transfer her allegiance from the Church of England to the church of Rome and promise the priest that she would raise any children to be Roman Catholics like their father.

And so, on that September day, Connie Sweetland, the man hired by Jack Williams to stow away his only son on a ship going to New York City; Connie's wife, Lotte, who later became Elrick Jr.'s godmother and whom we called "Nana"; Aunt Rose and Uncle Hugh Holness, the older cousins—brother and sister—who sponsored Nell's move to the United States; and, of course, Nell's cousin Mavis Holness banded together to witness this marriage, which Elrick had exacted from Nell—the same woman who said she never wanted a husband, who regularly went into and out of moods of despair, and who told her children that she wished they'd "never been born."

In very little time the more peaceful memories of the man "who bothered her" into marriage were erased by the violence that ensued.

PART THREE

GROWING UP WILLIAMS

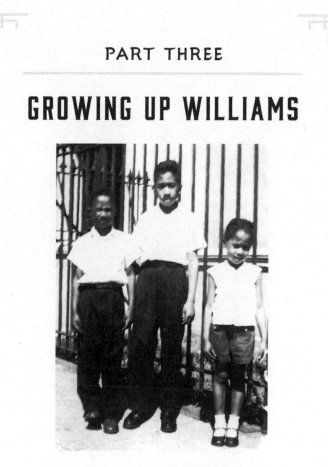

The faults of a superior person are like the sun and moon. They have their faults, and everyone sees them; they change and everyone looks up to them.

—CONFUCIUS

SCENES
FROM A MARRIAGE

IT MUST HAVE BEEN ONE OF THOSE GRAY, DARK NEW YORK OCTOBER DAYS IN
1947 when Elrick returned home, surly, maybe feeling a little guilty
when he saw his hugely pregnant wife, and asked where his clean
clothes for work were. Surely she gave him one of those looks that
said everything, that said your damn clothes are where you left them,
that said where the hell were you all night even though she knew. My
father was already wandering around at night, consorting with other
women, and sometimes not coming home. He was angry because my
mother had not done his laundry and his shift was about to begin. My
mother was furious, for other, more obvious reasons, and suggested
that if he wanted clean clothes, he could get them from the woman
who had kept him out all night.

How dare she? He was tired, possibly hungover, and not in the
mood for her razor-sharp tongue.

He slapped her. A single gesture that would demonstrate who was
in charge. There were more words between them. Then she turned
and walked away.

Nell walked down the hall of their apartment at 2089 Amsterdam
Avenue into the kitchen and opened a drawer. In it, she found the
rusty penknife she knew was there. While Elrick fussed in front of

the closet, she silently approached him and stabbed him three times in the back. She punctured his lungs. My father's first cousin, George, was in the front-bedroom apartment he had rented. He walked Elrick to Mother Cabrini Hospital on 163rd Street and Edgecombe Avenue, around the corner from our house. A few hours later two police officers came to the door to inform my mother that her husband had been jumped and stabbed by two men who had tried to rob him.

"Who told you that?" my mother asked.

"He did," one of them replied.

"Well, it's a lie," she said, thrusting out her hugely pregnant belly as if to underscore the point. "I stabbed him because he hit me, and if he hits me again I will stab him again."

My parents split up when I was about three years old, in 1955. The simple version of the conflict was that my father insisted that my mother had informed on him to the Immigration and Naturalization Service, telling them that he was an undocumented alien. My mother said that he was nuts, that the informer must have been a fat Jamaican woman named Winnie with whom he was having an affair. Winnie had both the motivation and the meanness to do such a thing, and was savvy about the workings of the INS. If she was angry at my father for one reason or another, a quick call to the authorities was sweet revenge.

My father had flown under the INS radar for a decade. Amazingly his luck had not run out until now.

Elrick Williams had entered the United States on the ship where he stowed away, got off in New York, and, like many immigrants before him, just stayed. He arrived, as my brother said, "without benefit of documentation." So he used a variation of Jamaican duppy-proofing—multiple identities—when the United States government became the evil spirit against which he needed protection. When he moved to the United States, Elrick Williams became Moncrief Powell. The name was not simply concocted from his fertile imagination. Moncrief Powell was his mother Sarah's first husband and was supposed to have resided occasionally in Cuba. Sarah and Moncrief never had children, so there was no namesake.

Moncrief Powell.

It was a useful name, dignified and aristocratic.

Mail would arrive at our home addressed to Moncrief Powell. Moncrief Powell had a social security number, a job, and three children named Elrick, Howard, and Paula *Powell*. At least, this was so for the early part of our lives. In the first grade I was oblivious of it. In the second grade I started wondering. In the third grade the confusion became more serious. I opened the principal's annual note to my mother, asking her at the start of the school year to clarify the confusion: "Is Paula Powell, Paula Williams?" My birth and medical records read "Powell," yet I was enrolled as and responded to Williams. By now I was used to the question, so I just stood in front of the school nurse who handed me the note to take to my mother. I read the note. "Yes, I'm Paula Powell," I said. "But my real name is Williams." And that ended the annual ritual of clarifying my identity with a note to Nell.

Let's just say that the INS was not inclined to forgive illegal aliens with assumed identities. When my father heard that the INS was looking for him—the neighborhood must have been alerted to the presence of the authorities, and the efficiency of our neighborhood communication system would have impressed Homeland Security— he was convinced that my mother's fingerprints were all over the summons. My mother retorted that there was absolutely nothing in such a strategy that would benefit her. Much as she had grown to detest my father, why would she have had the family's sole bread- winner deported? There was no reason for her to have done such a thing, but the same could not be said about that bitch Winnie. Mortal combat ensued. At one point my father sat on his bed tying his shoes and my mother, silent as a cat, approached him from behind with a kitchen knife.

And then she stabbed him. As she had years earlier.

My father was just returning from the hospital, with a paramedic on either side, when the INS agents took him into custody. He walked in with the paramedics and walked out with the agents. We didn't see him again for two years.

Combat was a regular feature of my parents' lives together. I was too young to have many vivid recollections, but my brother Elrick remembers the three of us sitting in chairs in the living room, a paralyzed peanut gallery watching the show, the regular feature of our father beating our mother. Howard said he wanted to become a cop so that he could arrest our father. My mother favored one of her legs for much of her life and couldn't stand very long on it because of the thrombosis caused by an injury. She said that in one of their more violent fights, my father kicked her and permanently damaged the veins in her leg. My father insisted that he didn't do it.

They could not help themselves, my mother and father; they were two strong-minded outside children, and neither would ever let the other in—except in the intimacy of combat.

GROWING UP WILLIAMS

SCENE ONE

ELRICK WAS FIRST, BORN IN NOVEMBER 1947. THEN CAME HOWARD, THE FIRST summer child, born in July 1950. I arrived two years later, in August 1952. My older brothers and I were born and raised on Amsterdam Avenue between 163rd and 164th Streets.

When we were growing up in the 1950s and 1960s, we were our own special club. We went fishing together in the Hudson River, or we would take the subway or ride bikes—I was usually riding side-saddle on the horizontal bar of Elrick's bike—north up to the Bronx and Van Cortlandt Park. Once we arrived we would get lost in the greenery. We took care of one another. We watched each other's back. More specifically, we were all responsible for one another, but my brothers were especially responsible for me, their baby sister.

Frequently, our mother didn't know the riskier details of our exploits. Sometimes we crossed the Metro North railroad tracks, careful—as we all knew from our "hoodrat" friends—not to touch the third rail and be electrocuted. Once when we squirmed through the broken chain-link fence with the "Danger" and "Warning" signs all over it, my shirt snagged and tore on a ragged link. When we got home hours later, my mother asked, "Paula, what happened to your shirt?"

"Oh, it tore when we were going through the fence to cross the railroad tracks."

Nell went to that place where mothers go when all vocalizations become unintelligible. She screamed something that was fast and angry and included the names Elrick and Howard. After she'd beaten the three of us with a strong, sturdy belt, we sat whimpering and promising never to do anything that dangerous again. I could handle the fact that my mother was angry, but my truth telling had violated something sacred among the three of us. For the only time in my life, I had broken solidarity with my brothers. They could not believe that I had told her, but lying wasn't something I knew how to do. I was six years old and proud to run the streets with my brothers and happy to share our exploits with our mother.

But I learned that day not to tell her about anything that was exhilarating. I knew then not to tell her about jumping across five-story rooftops, never to say anything about walking across the closed pedestrian bridge that spanned the Hudson River a few blocks from our apartment. I never told her that we would climb the fence and walk across Highbridge with its missing bricks and broken rails and roam that portion of the Bronx for about an hour before turning and retracing our steps back to our own neighborhood.

We learned independence; we sharpened our wiles; we acquired the skills of cursing and fighting. I learned that day to just *not tell* Ma the exciting details. And with Elrick and Howard frowning at me and threatening to leave me home, I never made that mistake again.

My mother let us know we were important as individuals but we were also the most valuable people on the planet to her, to our father, and to one another. She used to tell us that our poverty was an outcome of circumstances, and that these circumstance would end with our generation because we would succeed not only professionally but materially. She understood that there were times when professional success—becoming a great teacher, for example, or a humanitarian doctor—came without financial rewards. That was all well and good for others. But as far as she was concerned, not for us.

We went to Catholic schools and worked hard because our mother was not about to tolerate failure in anything. "You *will* be rich," she told us. It was our responsibility to compensate her for the suffering and the sacrifices that she endured.

Nell was the single most influential person in our lives, and she was a "fight to the death" kind of person. When I was about six or seven, she impressed on me the law of the neighborhood, if not of life. "If you get into a fight, do not leave until the other person cannot get up," she said. "Because you are *not* going to turn your back and have that other person get up and get you." For her, this lesson was not just theoretical or rhetorical.

We attended the Saint Rose of Lima Catholic school across the street and around the corner from our house. One day when I was about nine years old, I was walking home from school on 164th Street with another little girl. A twelve-year-old boy named Lionel who had an enormous head and a bad attitude was standing on his front stoop as we walked past. With no provocation—Lord knows what was going on in this kid's mind—he spit on me. I was shocked and outraged, nearly hysterical, as I turned and said, "You spit on me!"

He started laughing. "Yeah, I did." And kept laughing.

Disgusted and angry, we turned back. My friend went to her apartment and I went home crying and enraged. My mother was home—she was always home—and so were my brothers. My mother saw how upset I was and asked what had happened. Sobbing and furious, I told her. My mother got a certain cold look on her face and asked, "What did you do to him?"

I was astonished. This kid was three years older and easily thirty pounds heavier than I was. "What did *I* do to *him*?"

My mother slapped me across the face. "You are telling me that you let this boy spit on you and you're here? You didn't *do* anything to him?" At this point I was no longer crying because Lionel had spit on me; I was crying in disbelief because my mother had slapped me. She told me to get out of the house, go back around the corner, "and don't you come back—don't you dare come back in here until you tell me that you have beaten the hell out of this ruffian!"

Standing behind my mother were my brothers, who were eleven and fourteen. Out of my mother's field of vision, they motioned with their hands, waving for me to go out again, and pantomimed a reassurance that they would be there. I walked out the front door and they climbed through the window of the bedroom I shared with my mother and down the fire escape in the back and met me outside. They said, "Come on."

Like storm troopers my brothers and I marched around the corner to where Lionel still occupied his stoop. They ran up to him and growled, asking if he had spit on their sister. Lionel's bravado had faded fast.

"Yeah, well, I was only playing."

Elrick stepped forward, a terrifying fourteen-year-old. "You're gonna stand here and she's gonna beat the shit out of you," he said. "And you're gonna stand there and take it."

There was not much Lionel could say by way of providing an alternative plan.

I started punching him in the face and kicking him in his private parts and whaling him. All he could do was stand there and flinch. And every time he flinched, Elrick or Howard would bark at him, "Put your hands down!" I beat him until I was tired, all the frustration and rage from the indignity he had subjected me to provided extra energy. When I finished we turned around and my brothers retraced their path, climbing up the fire escape and back through the bedroom window. I walked in the front door, where my mother stood waiting. My clothes were a mess, my tears had dried, my fists were red.

"Did you beat him up?"

"Yes."

"Get in the tub," she said softly. She ran a bath for me and served me dinner.

I understood then the practical and immediate implications of one of my mother's rules for living: "There's no way in hell you are going to come back into this house and let me know that some kid beat you or bullied you. You will die first before that happens."

The corollary of this, which we found out later, was that if anyone ever tried to hurt any of us, my mother would kill him.

It was her way of expressing love.

NOCTURNE IN NELL MINOR

L ET ME SHOW YOU AROUND OUR APARTMENT ON AMSTERDAM AVENUE.
This was where I lived for my first thirteen years, and even today I can close my eyes and see the details: every crack in a wall, every worn corner on a couch, every drawer and shelf and utensil in the kitchen. My daddy lived there briefly, but I remember very little of that although my brothers remember it clearly. The apartment on Amsterdam Avenue was my mother's domain, and no matter how many boarders may have been transient residents, my brothers and I never considered it anything else. Though very humble, the apartment was always clean and tidy. It was as though Nell neatly organized the family's appearance to the outside world in a way that she could not organize our internal family relationships, much less her own relationship with our father.

We would walk up one step of the stoop, then up two more steps leading to a long white marble-tiled hallway with the front doors of two apartments. We lived on the ground floor, in a cookie cutout of many apartments in our neighborhood. Our five-story building was divided into eighteen apartments, and there were two on the first ground floor where we lived: Apartment A2, on the left, was ours and Apartment A1 belonged to Melvin and LaVerne Jones. We had a bedroom, a bathroom, and a kitchen—then the living room and another bedroom, which was the one I shared with my mother. These were the rooms in which we slept, studied, ate, fought, played, argued, talked, and sat silently with our mother.

My parents never shared a room; my father's room was the small bedroom in the front of the apartment. When my daddy moved out, a woman named Iris moved in and rented that space. Her most memorable trait was that she had a telephone—we had none—but hers was locked. When Miss Iris moved out, cousin George moved in and rented the room. For all those years, Elrick and Howard slept in the living room on a small convertible sofa. When Elrick was about twelve, he moved into our father's now vacant room, while Howard remained on the sofa.

The living room was the center of our household both literally and emotionally. It had a couch; a black-and-white television; an armchair; a coffee table, the kind with a brown leather inset bordered with a gold-leaf line; a matching side table; and walls that were usually painted either eggshell or soft green or pale sky blue. Nell loved to make the wall colors soft, never vivid. There was always an oil painting in a large frame hung from a silk rope contraption. On another wall hung a Chinese silk screen with bamboo-slat backing, the kind you can roll up into an articulated bundle. Sometimes an outdated pictorial calendar or a lovely pen-and-ink drawing hung in that space, reminding our mother of China, which she never saw but to which she felt connected.

The living room was also the major thoroughfare of the apartment because in order to get from the front to the back, to the bathroom or to the kitchen, or out the front door, we had to go through the living room. After Elrick moved into our father's bedroom, Howard slept alone on the living room couch. He seemed to be able to sleep through anything—which was fortunate, because while the rest of us were early risers, Howard could be comatose until noon.

There were two times of day when I had our mother to myself:

at bedtime and from four a.m. until five a.m. We shared a room and a bed; her room was my room and her bed was my bed. I would turn on my side facing right, swing my top leg onto her sleeping body, and drift off to sleep. In truth, Nell was up long before I roused myself at four a.m. She would rarely go to bed at the same time I did, so during those few hours when she actually came into the room and slept in our bed, I would nestle my body into her warm embrace and swing my leg over her. It's strange that I remember childhood sleep as being in my mother's embrace, yet she almost never embraced me when we were vertical.

Often I would awaken to find her sitting at the small table in the kitchen sipping a cup of Lipton tea. She would leave the tea bag and the spoon she used to stir the tea inside the cup, and add boiling water, sugar, and canned evaporated milk. After she made her tea, she would put the bag in her mouth and suck out the rest of the flavor.

I'd wake up and pad into the kitchen, where she sat at the table with the three chairs around it. Smiling, she would begin to prepare my morning drink in my special cup: warm evaporated or canned milk, mixed with water, a small pat of butter, and a spoonful of sugar. Heaven! Most of our dinnerware was mismatched, as was the sterling silver cutlery. Sometime in her past, she had assembled nice tableware, but what remained were shards of her earlier hopes and dreams. From among these remnants, I had "my" cup, fork, spoon, and plate.

By the time I met her in our kitchen for our predawn quiet moments, she had already retrieved my brothers' white uniform shirts and my blouse from the refrigerator. After having hand-laundered them on a washboard the day before, she would suspend them on hangers over the claw-foot bathtub to dry. Then she would sprinkle them with water, roll them up tightly, encase them in a plastic bag, and leave them on the refrigerator's lowest shelf for a few hours. Only then were they ready for the iron and ironing board. That was how the Williams kids went to the Saint Rose of Lima School each morning: with startlingly bright, crisply pressed white shirts.

Those few hours before dawn, before school, were times of special closeness between my mother and me. I would sip my warm milk and watch her iron my brothers' shirts and my blouse and refresh the uniform slacks, ties, and jumper. Ma would use this time to teach me about ironing, cooking, and sewing—all skills she said I needed. "You have to know how to do these things," she said, "so that you have good meals, are dressed well, and have a clean house. But you will grow up to *hire* people to do these things for you. *Do you understand?* You *will* have a clean bathroom and toilet but you *will not* spend your time making them so. Earn enough money so that someone else does this drudgery. Not you!"

That was what Nell wanted for her only daughter—not a husband, not marriage, not children. Her firmest conviction, and her major goal, was that I was not going to have a life like hers.

Nell's daughter would be a professor or a doctor or a businesswoman. Had I told her I wanted to become a lawyer, she would have pulled my tongue from my head, as she sometimes threatened to do. "Lawyers are liars. They should just call them that," she said. "They are paid to lie and I will not have my daughter in such a dishonorable profession." End of conversation. No way was her daughter going down that path. I never understood what had soured her, but to Nell lawyering was not a career option for any of her children.

The early hours were also a time of confidences, when the defenses with which my mother armed herself for the daytime were lowered. She told me that her middle name, Vera, came from her godmother, a Cuban woman named Vera who was her mother's best friend. One morning she told me about being raped when she was twelve. She revealed this to show me the untrustworthy, bestial nature of boys and men but also to share with me an aspect of her life that she could tell no one else.

Another morning she told me that her grandmother often addressed her as "you half-Chiney wretch." A recurring theme during those hours was that she never knew the love of a father and she never had a relationship with her mother. Sometimes, when she was feeling deeply depressed, she would say that she should never have had children—or that she should have married "the Chinese man who wanted to marry me, and not your father."

Slowly the rest of the house would wake up. I would hear music in Elrick's room as he studied or read before school. Eventually Howard would rouse himself and get organized for the school day, which was a torment for him. And the time of solitude and confidences between me and my mother would end.

Until the next morning.

FINDING MY OWN SPACE

I AM TEN YEARS OLD AND IT'S A HOT, TRANQUIL SUMMER MORNING IN HARLEM. I have no real plans except to get outside as soon as possible. After I eat breakfast—my mother divides the word into a verb and a noun, "break fast"—I escape the apartment. "Going out to play," I tell my mother. But I am not going out to play. I am going to meditate, to renew myself, to ponder the world, to be alone.

I start walking uptown and reach Edgecombe Avenue and 166th Street. My destination is my favorite rock in a wooded area of Edgecombe Park. The park isn't as large as Central Park, but it is sprawling, with baseball diamonds, bocce and basketball courts, bike paths, woods, nice rocks to climb, and a fountain to give it some gravitas. My special rock is large, with a flat striated surface and is warm to the touch at the height of summer because the sun beats down on it most of the day.

I am by myself, and my mother would probably be furious if she knew that I had gone into the woods all alone, but I love the solitude and the sense of autonomy here on my rock, my private property. I sit like an Eastern deity, with my knees crossed and my eyes closed, facing the sky. The color I see behind my eyelids is red, and I'm soothed. This is a retreat I can access anytime—the world behind my eyelids—but only when I am at a place like my rock does the warmth of the sun turn the world behind my eyes a warm blood-red.

And then I start daydreaming about my Chinese grandfather. I wonder where he is. Sometimes I imagine that if I concentrated all my mental and emotional energy on conjuring his physical presence, I would be able to talk to him and ask him where he has been, why he has never given us a sign of interest or concern, why he disappeared, and—most important—why he abandoned my mother without leaving even sliver of evidence that he had existed at all.

I would often sit on my rock, serene and secluded. To achieve that state, I needed only to go a few blocks uptown. My mother, in contrast, even in the busy streets of Harlem, was always alone, always unconnected. On Amsterdam Avenue, she had few real friends, no husband to love her, no mother to counsel her. Her only hope for a better life rested with the three of us.

I would think about the rest of our family in New York. My mother's much older cousins, Uncle Hugh and Aunt Rose Holness, were Jamaicans of a quaint bygone era, who had moved to Harlem earlier and who offered my mother her first home when she arrived. Uncle Hugh wore a cracked, broken-brimmed woolen felt fedora even in the August heat. Aunt Rose's thick flesh-colored stockings were rolled to her calves, and secured by a little knot just below her skirt. They moved around their dark Harlem brownstones in their too-old and too-worn house shoes. Their clothes were relics of styles from a distant past, clean but lacking any color or shape after many scrubbings on a washboard. The linoleum floors were faded and well trodden, and had cracked in so many places that other strange-colored floor coverings peeked up from below. How many years was it since anyone had seen the real floor beneath all that?

We were always dressed up when it was time to visit them and were told to behave. But I wondered: Is *this* where my mother and I come from? Are these two stern elderly people what "having a family" means?

I didn't want to go there. I was much happier on my rock, pondering the mysteries of my grandfather's life and my mother's Chinese background. Cracked felt fedoras and flesh-colored stockings I knew. I was surrounded by them in Harlem. But China was very different. China was mysterious and distant—a place of silks and tea, communists and revolutions, savory foods, a language that sounded like birdsong, evocative pictures of men wearing little jackets that looked like dresses.

"He went back to China and died." Each part of that sentence was so thought-provoking that I needed to get away, go to my rock, close my eyes, see the colors, and try to wrap my mind around it.

Then, with a start, I would come to my senses. The sun was higher and hotter. My mother would be wondering where I was. It must be close to lunchtime. And my thoughts would return home, to Elrick and Howard, who might take me fishing that summer afternoon.

UMBILICAL CORDS

I T WAS PREDICTABLE THAT MY BROTHERS AND I WOULD BECOME NELL'S ENTIRE life. We were all she had that was truly and completely hers; and no matter where we roamed or with whom we may have been, that fact never changed. It is tempting to say that she never was able to sever the umbilical cord.

This is, of course, a familiar metaphor. But in our case it is not just a metaphor. Sometimes my mother would open her dark brown leather keepsake box. It was a hard-sided rectangular purse with an art deco gold-tone clasp. I lived for the moments when we would sit on our bed and she would pull out the box, lift the clasp, and remove a manila envelope just a few inches square. Then she would unwrap a white cotton ball that covered some dried, purple-dyed, gnarled, clumped matter. I would look at the shriveled bits that stuck to the cotton and think that they looked like eggplant-colored beef jerky.

"Mommy, what are those?" I asked the first time, fascinated by these grotesque bits, kept with such care in a special place reserved for her most precious possessions.

"This," my mother said with some pride, "is your umbilical cord. When you were inside me, this is how my body fed you until you were born."

"Why did you keep them?" I asked, still fascinated but now slightly repelled by this strange treasure. Was it something that every mother kept? I doubted that. My mother would always do things other mothers wouldn't imagine doing. And even for her, this seemed strange.

"Someday," she said, "I am going to bury them under a tree on a property somewhere that I will own, that you children will buy me." The property would be near water, we thought, and would have a luxuriant garden.

There might have been many reasons for her to save those bits of the umbilical cords that had tethered us to her—for example, an old midwife tradition in Jamaica involves burying them—but I am sure she had her own special reason. These bits of dried tissue were proof beyond all doubt that we were *her* biological children, that we came from her body. No questions need be asked. Elrick, Howard, and I were not outside children. We had no unregistered birth certificates. We knew we came from Nell and Elrick. We may not have looked much like her, but that was irrelevant. She was our repository of personal documents: she had our birth certificates, our baptismal certificates, her marriage certificate, her citizenship order, our umbilical cords, and her passport—all in her keepsake box.

When my father was deported to Jamaica in 1953, my mother started the process of filing yet another round of documents to try to get him home. My father had sent onionskin *par avion* letters from Jamaica to 2089 Amsterdam Avenue. In the series of letters that he wrote from his mother Sarah's house, he pleaded with Nell, "If you become a U.S. citizen, then I can come back to live with you and the children." I was just a year old when he was deported and would be three when he returned. My mother wondered: Was history repeating itself in our family? Was there some cursed cycle in which *she* would be partly responsible if her daughter grew up without a father? That irony was too painful to contemplate.

So Nell went to Catholic Charities for help. She had been raised as an Anglican in the Church of England but Elrick had been raised Roman Catholic. My mother converted and vowed to raise as Catholic any children they might have. The appeal to Catholic Charities was my father's inspiration. He figured that the Catholic Charities near Harlem would be used to such situations and could advise Nell how to get him back, this time *with* "benefit of documentation." The caseworker listened sympathetically. She must have heard hundreds of similar stories, and she said that Nell first had to become a United States citizen herself.

Catholic Charities helped her file the necessary paperwork, and she supplied the witnesses. One witness was the superintendent of our building, who—with his wife—was like family to us. The other was a friend of my father's who had a good, steady job. They signed the documents and attested to her good behavior, admirable character, and citizenworthy commitment to the United States. On June 6, 1955, the eleventh anniversary of D-day, she was granted U.S. citizenship in the Southern District Court of New York. Not long afterward, my father legally returned to New York and to his family.

Togetherness didn't last long. After several battle-scarred months, their estrangement became permanent. When I was three years old, my father left my mother, my brothers, and me. He moved out for good.

CUSTODY MATTERS

WHEN OUR DADDY, AFTER RETURNING FROM HIS JAMAICAN EXILE, MOVED out of our apartment, my parents' official divorce was still many years away. During the ensuing custody battles, my mother made sure to point out our father's love for us. "You know your father wants you," she would say, and this was a compliment—albeit a rare one—to my father, as well as an affirmation of our worth. Worth was a major theme in our family. Our mother wanted us to grow up knowing that our lack of material comforts was not in any way a reflection of our inherent worth. By way of emphasizing our worth—and I suppose mine in particular, because I was the only girl—she frequently reminded my two brothers that, unless they wanted to incur a beating, they had to take care of their baby sister, and of each other.

Yet, as any child of a broken home is likely to know, such expressions of love from both parents do not exactly make the child's life easy or comfortable. In our case, the question of who was going to have custody, in a relationship that was already toxic, raised the conflicts to even higher levels of intensity. Elrick, Howard, and I were not consulted as to our preferences. We loved our daddy, but our mother's humble Harlem apartment was home.

Our father was determined to create a new home for us, with him, so he bought a house in Springfield Gardens, Queens. The neighborhood was a seventy-five-minute subway ride from Harlem but it had so little in common with the place we knew so well that it could have been in another country. We had no friends there. We had no connection to the place. Mostly white folks lived in Springfield Gardens. We would go there, and sometimes we would find that one of our father's new companions had taken up residence, perhaps with her child. These companions' responses to us ranged from mere dislike to active contempt. We reciprocated, but were powerless.

Finally the issue of custody went to court and we were all called into the judge's chambers. I was about six years old, and this was a time when a judge in Family Court would see no harm in assuming a fatherly attitude and asking a little girl to sit on his lap as he talked to her about her family situation. At the time, it was a gesture designed to create an atmosphere of safety, candor, and confidences.

My brothers looked on with suspicion and some amazement. I was used to getting special treatment and they were used to seeing me get it; but watching their little sister sit on the lap of a black-robed white judge, as if he were Santa Claus, was more than they had been prepared for, even though they were aware of my exalted status. We spoke a bit about things I cannot recall, and then the judge got down to business. "Honey, you know that both of your parents would really love you and your brothers to live with them," he said. "But of course you can't be in two places at once."

I nodded. I looked at my brothers, who were standing next to me and the judge. This judge understood the problem.

We had spent the weekend at our father's house, and en route to the courthouse, he had prepped us. "If the judge asks you where you want to live, you tell him you want to live with me," Daddy said sternly. "Do you understand what you are to say?" We nodded silently, but my inner voice screamed, "I don't want to live with you. I don't want to live with you. I want to go home with my mommy!"

Tears spilled from my eyes and I could also see tears forming in my brothers' stoically narrowed eyes.

It all led me to this moment of truth, sitting on this white man's lap when he asked me the question that our father said had to be answered one way: that we wanted to live with him in Springfield Gardens, Queens, and not with our Ma in our home in Harlem.

The judge continued. "I can make the decision of where you will go. But I would really like to get a little help in making that decision. Where would you like to live?"

I glanced briefly at my brothers. We all knew where we wanted to be and we all knew that I was not going to lie. My daddy may not have known that I hadn't developed an ability to tell white lies, to be tactful, to please an authority figure. My brothers knew. The three of us sure—as sure as we were of anything—that I would say what I wanted and what I knew we all wanted.

"We want to stay with our mother," I said. "We want to stay in Harlem and not live in Queens. Our father told us to tell you we want to live with him, but we don't. We hate it there and we just want to go to our mother." My brothers gave me just the slightest smiles. We were a trio, joined in ways that we sometimes can't explain even today. Life is a war; we are a band of soldiers; we stick together.

I looked at the judge. He nodded gently. I don't think that this came as a surprise to him. "Well then," he said. "I will see what I can do."

That day, we took the train from Queens Plaza back to Manhattan. Back home to Harlem. Back to Nell's care. My father was outraged by his defeat and months passed before we saw him again.

GROWING UP WILLIAMS

SCENE TWO

ONCE MY BROTHER ELRICK GOT INTO A FIGHT WITH A KID CALLED FREE-ZEE: whether that was his real name or even the correct spelling I will never know. Free-Zee came from a very large family living a few blocks away from us. He and Elrick were playing basketball and one thing led to another and a fight started. My brother won and, as a parting shot, picked up a pair of sneakers Free-Zee had left behind and brought them home. He wasn't going to wear them; they were just souvenirs of war.

Then one of the local town criers—a kid who liked to let everyone know what was going on—ran to our apartment and announced that Free-Zee was coming to the building with his father and brothers to reclaim his sneakers.

"These people are coming here?" my mother asked. That was enough for her. This moment aroused all her survival instincts. She had grown up without a father and she would be damned if she let her children feel unprotected just because their father was not around. She would be the man. Better still, she would be a lioness.

First, she needed to arm her troops. She picked up her supersharp meat cleaver, which she honed on a square stone block almost every day. Knives and scissors in our house were never dull. And since Nell had used her sharp weapons on her own husband—twice—strangers would be easy. Having retrieved the cleaver from the stand where it was kept, she directed Elrick to get his baseball bat and dispatched Howard to find the old pair of handcuffs, which dated from World War I and weighed about seven pounds. They had belonged to Elrick's godfather, Conrad Sweetland, who gave them to his godson. On this occasion they were to be used again for warfare, not as handcuffs but rather as a weapon that was almost medieval; by holding one end and whirling the handcuffs overhead you could knock someone senseless. We also had a cat named Frisky, who was usually gentle and sweet but became ferocious if she was tossed, at which point she would dig her claws into whatever or whoever was within her reach. "Bring Frisky," my mother ordered. Frisky was my weapon.

When we were all armed, we walked out of our apartment to the stoop. As she stood there, my mother leaned against the building—she often did this, but now she had the meat cleaver in her hand, hidden behind her back. I felt that Frisky and I were part of an unbeatable team. The enemy approached: a large, well-built father with his four sons. The neighbors gathered. They knew that my mother and the three of us were not people to be messed with.

The father came forward, ignored Nell, pointed to Elrick, and asked his son, "Is that the kid?" His son nodded. He came farther forward to start disciplining a kid he considered both defenseless and a punk, but before he could get a word out of his mouth, my mother grabbed the front of his shirt and delicately placed the cleaver at his jugular vein, pressing ever so slightly so as not to break the skin.

Yet.

"Why are you here?" she asked. "What do you want?"

The man's bluster and menace disappeared. The dynamics had

completely changed, thanks to a 120-pound half-Chinese woman with a meat cleaver and an attitude.

"I came to get my son's sneakers," the man said quietly.

"Not like this you didn't," she said, pointing her chin at his other sons. "You didn't come to get his sneakers. You came here because you thought you were going to beat up my son."

There was a long pause. The man stood stock-still. Tears began to leak from his eyes, and he barely breathed.

"I will kill you while you stand here."

Everybody knew that no one should make a move.

"Now, I'm going to let you go. Tell your sons to go back."

They went back.

"Elrick, give him his shoes."

Elrick tossed the shoes to Free-Zee.

"Now, here's what I want you to know," my mother said. "You don't live in this neighborhood. You don't live on this block. Don't you ever come to my house again. My son will not play with your son. Your son will not play with my son. There will be no fights. But don't you ever come over here again."

With that, Nell Vera Lowe Williams actually pushed him away, and then she just stood there with the cleaver. Chastened, and not a little humiliated, Free-Zee's father turned and walked away with his sons.

The whole neighborhood stood in quiet admiration and fear of Miss Nell. Gradually a few teenagers went back to playing stickball; three girls took out their jump rope and began playing double Dutch; a mother walked up the stairs to her apartment; a few kids started walking toward a neighborhood store. Slowly they all returned to whatever business had been interrupted by our own Nell and Goliath story.

Had anything else happened, had my mother not defended us, had the man managed to get the upper hand, then the three of us would have become easy targets for nearly everyone in the neighborhood. It would have been open season on the Williams kids. For my mother, the audience was as important as the rest of the show. She was glad everyone was there to see her in action. "I'm going to show you and everybody standing here," her actions said. "If you come after my kids I will fucking kill you, right where you stand."

BREAKING MY DADDY'S
HEART AFTER HE BROKE MINE

AFTER MY BROTHERS AND I ENSURED THAT WE WOULD NEVER HAVE TO LIVE with our father again in "Siberia," he disappeared from our lives for a while. We'd reconnect and then he would exercise his court-ordered visitation rights. We would spend some weekends in Springfield Gardens, Queens, and in summers we'd endure some weeklong imprisonments.

Our father was furious because we wouldn't live permanently with him in the nice house that he said he had bought just for us. Not only did we not want to be exiled in a mostly white neighborhood in Queens where we had no friends, we didn't like Daddy's live-in woman, Miss Alice, or her disgusting son Pat—who once, for just a few seconds, molested me until I pushed him away.

On the other hand, the two of them didn't like us much either.

The years passed. We all graduated from grammar school, and I exchanged my blue plaid uniform for the navy blue uniform of Cardinal Spellman High School. Although we spent some time with our father, our center of gravity was our mother. But just before I was to turn thirteen, I discovered that my grandmother Sarah had been staying with him in Springfield Gardens for nearly the entire summer. I had never met either of my grandmas, and for a twelve-year-old girl with a strong sense of family, this was the final blow.

When I heard from my cousin George about Sarah's visit, I sent my father the most vitriolic letter I have ever written in my entire life. All my rage at him and disappointment in him poured onto the page. How dare he deprive us of our ancestors? How dare he bring a family member into this country and not have the simple decency to swallow his pride and introduce his mother to me? No matter that my brothers remembered our grandmother with only dislike. I deserved the chance to get to know her and decide for myself if I disliked her. After all, there was a remote possibility that I might actually like the woman.

A few days after I mailed the letter, our father drove his long coral car with the white roof and boomerang taillights to our neighborhood. As was the custom, one of the neighborhood kids spotted him and appeared breathless at our door, saying, "Miss Nell, their father is coming around the corner!" My brother Elrick wasn't home but Howard, now fourteen years old, and I were there with our mother.

Howard wanted nothing to do with our father. "I don't want to see him and I'm not going to see him!" he said, defiantly. Our mother was not going to stand for this foolishness. She stood the two of us side by side. "He can't kill you," she said, implying that a risk of death was the *only* circumstance in which avoiding someone would make sense. "You will go with your sister right now and you will tell him exactly what you want to tell him. Go now. And you *will* be there with your sister."

Howard was furious with me and anxious about having to act as the eldest in this situation. It was Elrick Jr.'s role to be the general of our small battalion. Howard did not want to be in charge. I couldn't have cared less. I didn't need Howard. I was ready to fight and I was going to take my father apart. Outraged, indignant, and wronged, like a heat-seeking missile, I targeted my father for destruction.

We met him as he was walking on Amsterdam Avenue, just before he reached our building. He took my hand and walked us to Edge-combe Avenue and 163rd Street, where his car was parked. He opened the door for Howard and me to enter and sit on the back bench seat. And, oh my God, in the front passenger seat was Miss Julie Jackson, his current girlfriend. I hadn't seen her since he threw out Miss Alice and her creepy son, Pat.

Once, when my father and I were alone, I told him how Miss Alice mistreated us when he wasn't around. I told him how she forced Howard, the gentlest and most loving of the three of us, to wash pots in a dark cold basement all alone after my father fell asleep. He was stunned and asked me why we hadn't told him sooner. "I thought you *knew* what she was doing to us," I said. He gasped and said he would never let anyone hurt us. That was when I loved my father most of all. He evicted that woman who had mistreated us; he slew the monster—and her monstrous son—who had terrorized my brothers and me. He knew only too well the experience of being in a house with a violent older woman; his memories of May had prob-ably faded a bit with time, but the image of his own flesh and blood in a similar situation, when he actually had the power to help, must have been devastating.

Now, however, it was time for my father to confuse our lives again. I boldly leaned my head into the front of the car and saw that Miss Julie was pregnant! Pregnant? My head exploded and, as has often happened to me, I became two Paulas: in this case, one sit-ting in the car and the other floating above, observing and record-ing the scene to be replayed over and over for the rest of my life. My father started his well-intentioned monologue with an apology, a futile attempt to make things right with me. He admitted that he was wrong; he said it was wrong for him not to bring my grandmother and her three grandchildren together. In fact, he was going to make amends and take me to Jamaica in a few weeks to meet her.

The words just fell around me like bits of trash carried by the wind.

"Is that woman pregnant?" I demanded, sotto voce. I must have sounded bone-chillingly throaty. "Is that *your* child?"

He paused as Julie began sobbing softly. "Yes," he said.

I could not contain my rage. "How dare you? How *dare* you? My brothers and I are eating potatoes every day because they're only five cents a pound. My mother stays up nights, darning and repairing customers' clothes for Mr. Chung's laundry to make money to feed us, and you're knocking up some woman, driving the latest long car, and taking care of *her*? How *dare* you? You're my father; you have three children who don't have enough! *We don't have enough!* And you dare to take what's ours and give it to this woman and your bastard baby?"

I was on a roll. My twelve-year-old self could not contain my outrage. "Do you know that I sometimes go hungry? Do you know that we don't have enough food to eat sometimes? Do you know that my mother eats last and sometimes, to make sure we're fed, she doesn't eat at all?"

I ignored Miss Julie's tears and turned on her. "And *you're* going to marry him? Oh, but you can't marry him because he's still married to my mother. Did you know that?" She looked shocked, but she had no time to ask my father if this was true before I turned on him again.

"What kind of man are you that you would bring your mother here? You know that I have never laid eyes on my grandmother, and whether you're mad at me or not, you kept me from her."

Miss Julie's tears were now accompanied by full-throated sobs. Her shoulders heaved and she turned toward me to say something.

"Don't speak to me!" I shouted. "Don't you say a word to me. You're a wicked, wicked woman and he's a wicked, wicked man!"

I broke my father's heart in record time.

He deserves it, the floating Paula told the seated Paula. Don't back down. He can't be your father. Your daddy would never treat you this way. His birthday is August 25. My birthday is August 24.

"You're the best birthday present I've ever received" was my daddy's quiet refrain to me.

His chest was against the steering wheel.

"P," he said. "P, I'm so sorry. I'm here; I'm here and I won't go away again."

I wanted to believe him. I wanted to trust him. I wanted to sit on my daddy's lap and have him hug me as he used to. Instead, I said, "I will never, ever forgive you for this." I said, "We have no relatives anyway. I've never met any of my grandmothers or my grandfathers."

My father must have seen an opening. "P, I'm sorry, and I want to take you to Jamaica to meet your grandmother."

"You want to take me to Jamaica?" I said. "She was here. Now you want to take me there?"

"You're right," he said. "You should have seen her. She should have met you. You should have met her."

Howard opened his door and practically leaped to the sidewalk. To say he was stunned by what he'd witnessed—without uttering a word—would be an understatement. He'd been worried that he'd have to defend me from an angry daddy. He hadn't anticipated that his baby sister, born in the year of the dragon, would or could be a wrathful avenger.

Our father got out and opened the rear driver's-side door for me. I hurried around the car to the sidewalk and he hugged me. It wasn't easy. It isn't easy to hold a wrathful avenger close. I stepped away, glared at him over my eyeglasses, and tilted my head defiantly.

Julie sobbed. Howard watched. Daddy pleaded. I turned and walked away.

Daddy said he promised: he and I were going to Jamaica in a few weeks.

As Howard and I walked past Mother Cabrini Hospital and then the rest of the way down 163rd Street, I replayed the scene that the floating Paula had captured. Everything around us seemed eerily silent to me. The traffic on the summer streets, the shouts of kids playing stickball, the rhythmic slap of the jump rope hitting the side-walk as girls played double Dutch, the Sounds of Motown blaring from a record shop—I heard none of it.

We walked into our mom's lair, where all was safe. In her emotional state, she reverted to Jamaican patois. I heard her ask, "Tell mi 'nuh! Whatappen?"

Howard came to life. He was animated. He told Ma that P said this to Daddy! And she told him how he's a bad father! Then she told Miss Julie that she was a wicked woman, taking him from us and having a bastard baby!

What? My parents were still legally married. There never had been a divorce; not even a request from him for a divorce.

"*Pregnant?*" Was Nell reeling from this? I know my mother remembered their long-ago vow that they would never have a child out of wedlock. They would have no outside children. Another promise broken. Another standard violated.

My older brother Elrick came home during one of the hundred retellings of my confrontation with Daddy. He listened. He chuckled.

Cousin George came the next day. He smiled and said he was proud of me. He said my father needed to hear that from me. He had cash with him that he said my father asked him to deliver to my mother. Daddy would no longer withhold the child support payments to punish us for not wanting to live with him. The years of alienation from our father were over.

I MEET MY
GRANDMA SARAH

THE BRITISH COMMONWEALTH GAMES—A KIND OF OLYMPICS—WERE HELD IN Kingston, Jamaica, from August 4 until August 13, 1966. There were thirty-four teams from the Commonwealth, competing in swimming, rowing, boxing, badminton, and so on. The island was in a state of excitement, but for us the games were mostly background noise; we had family to meet.

My father had kept his promise.

This trip was his way of trying to compensate me for several years of alienation and his pregnant girlfriend. He booked first-class seats to Jamaica on Pan Am. It was my first flight. I was all dressed up and had some presents that my mother wanted to send to the remnants of her family there. One in particular was for her sister, my aunt Hyacinth. It was a Longines-Wittnauer watch, with tiny diamonds on both sides and a thin black cord as the watchband. I have no idea how my mother managed to save the money for it, but it was a precious and important gift. My mother made me swear not to tell anyone about the watch and to be as careful with it as I could possibly be.

My father worked at Valley Forge Steel building airplane engine parts. As we settled into our luxurious seats, he talked about the functionality of the engines and how, although planes are heavier than air, the engines made it possible to take off and become airborne. His conversation brought back memories of going with him to Floyd Bennett Field in Brooklyn to watch the air shows when I was three years old.

Planes would buzz overhead at Floyd Bennett as Daddy told me and my big brothers all about aeronautics and aerodynamics. His passion and enthusiasm were infectious. My brothers, encouraged by our mother, spent hours in our small apartment on Amsterdam Avenue with balsa wood, glue, and single-edge razor blades, constructing, deconstructing, and reconstructing airplanes. We'd pinch pennies for materials and my brothers would buy small propeller-driven engines at the hobby shop. After we completed the projects, we would head over to Edgecombe Park with a phalanx of other kids behind us, and with our ever-vigilant mother and our dog, Duke, to see if these engineering experiments would actually become engineering feats.

My daddy wanted to be sure that my first real flight would make me love flying. It did. The stewardess came to take our meal orders. My first airborne meal was served in a large, exquisite, clean, white clamshell. The inside was pearlized and contained small scallops in a cream sauce. And, oh my God, it had mashed potatoes piped around the rim! The entire concoction had been put under a broiler or fired somehow, because those potatoes had a singed brown crusty finish. I smiled to myself at the luxury.

Each tray bore a white linen place mat. And on each tray there was a set of salt and pepper shakers about 1.5 inches high, clear glass with a silver-plated screw top. I will never know if the stewardess who removed our trays realized that the shakers were missing.

Daddy had given me Amy Vanderbilt's book of etiquette when I was six and I had read it and committed some of the most important points to memory. By the time I was seven, I knew how to set a formal table for an eight-course meal—an important skill to have in our two-bedroom tenement in Harlem. Daddy wanted each place setting in his house in Queens to have its own salt and pepper shakers, and Pan Am helped him achieve his goal.

Much as I loved the experience, I could not stop thinking of the times at our Harlem apartment when we didn't have enough money for the basics. Yet Daddy owned a house in Springfield Gardens—he put Elrick, Howard, and me on the deed so we would grow up knowing we were homeowners—and he could fly me first class to Jamaica. I was impressed with the luxury, the finery, and the menu but was resentful at the same time.

When we landed in Kingston, I suddenly met some family. My aunt Ouida, who was my father's stepsister, and her younger brothers—my uncles Paul and Maxime—were there. Uncle Max worked in customs and was at the airport waiting for us. These were Marques Harrison's children, Grandma Sarah's stepchildren, Daddy and Aunt Cutie's stepsister and stepbrothers, and Uncle Harry's half sister and half brothers. But in Jamaica, there's really no "step," no "half." You're family. They embraced me warmly and lovingly. And once my father was in Jamaica, the land of his birth, the place of his citizenship, he became even more formidable. He also became even more Jamaican, with a more pronounced accent; he used more patois than I'd ever heard him speak; and his laughter had more of a lilt. My father changed in front of my eyes. He was even more confident, even more in charge and at home, than I had ever seen him.

I breathed in the hot, humid air. I almost felt as if I were drinking the rich fragrances, a heady mixture that can be found nowhere else but this island. It was sharp, with the smell of the ocean and of bodies sweating, sweet like sugarcane, subtle like a curry. I tried to look everywhere at once. For the first time in my life, the lilting Jamaican patois, which emphasized how alien my parents were in Harlem, surrounded me from every direction. I didn't need to translate what I heard, but my father sometimes had to translate what I was saying.

My father then brought me to meet my grandmother, his mother, Aunt Sarah.

My mother had warned me about Aunt Sarah, who may have been responsible for bringing my parents together, but who had long since betrayed, or hurt, my mother in ways Nell would never forgive.

We finally arrived at Sarah's house, 4 Melrose Avenue, a rambling colonial plantation house with a wide veranda and jalousie windows such as I'd never seen in Harlem. My grandmother was brown and short and had the plump breasts and bulging waistline that were supposedly typical of grandmothers. Her hair was gray, undyed, and unpressed. It was braided into two crinkly, curly coils held at the sides of her head by hairpins. Her house was full of knickknacks and the chairs had starched lace doilies on the seat backs and armrests. Mahogany was everywhere. Ceiling fans stirred up a gentle yet hot and humid breeze. Sarah shared the house with her husband— Marques Harrison—and the household help, who lived in a room outside in the back. I could see that my father was proud to show me off, and a bit wary around his mother. I asked him where my mother's sister and mother lived and he said they were about a mile, maybe a mile and a half, away. Not a great distance at all. We could have walked there.

But for three days we didn't go.

I started pressing. His mother made no effort to take me to meet these long-discarded in-laws. My father was too busy reconnecting with, and showing me off to, relatives and friends around Kingston. We went to Uncle Peter's furniture shop on Beeston Street. We visited a newspaper, the *Daily Gleaner*, where my dad's friend was editor, and this nice man made me a metal stamp of my name, Paula Williams, from the old hot-type press machine that they still used. We ate meat patties and drank fresh water from a coconut that had been broken open with one swift swipe of a machete. I loved Kingston, but I was still lacking an important meeting. "Daddy, when am I going to go see my grandma and Aunt Hyacinth?"

"I'm going to take you tomorrow," he said.

That evening I prepared for the big reunion. I chose my clothes and laid them out. I pulled out the gifts that my mother had so carefully selected, including the watch that I had hidden for safekeeping. But when I opened the box and looked inside, the watch that my mother had packed, the precious Longines-Wittnauer, was not there. In its place was a cheap, old, broken Timex. I could not believe my eyes. Oh, my God! This was not the watch my mother had given me. In a panic, I called for my father.

My father came into my room and saw how upset I was.

"What happened?"

"Mommy gave me a watch to give to Aunt Hyacinth," I said, gasping with shock. "But that watch is gone. This is not the watch." I was nearly in tears. This was the most important errand of my life in Jamaica, and it was all falling apart. How could I possibly explain this to my mother?

My father's jaw tightened. "Give it to me," he said. I handed the gift box of junk to him. He said, "Wait here."

Moments later, I heard my father screaming at his mother, hurling rage and abuse at her. He was used to her nefarious schemes, but he could not believe that she would visit them on the granddaughter she had only just met. He came back with the Longines-Wittnauer watch, and I just looked at him. I was glad that he had protected me, but I was stunned that he had to protect me from my own grandmother. He said only, "I'll take you tomorrow to your grandmother and your aunt Hyacinth. The watch is safe."

I met Grandma Albertha, Aunt Hyacinth, and Hyacinth's only child, cousin Norman, the next day. Albertha seemed distant, or perhaps she was just old. My inner self asked her: Why did you take my mother away from her father? Why didn't my mother grow up with you? Why don't I feel any warmth or love from you or for you? Why are you alive and why is my grandfather dead?

But my mother's half sister, my aunt Hyacinth, was a full Jamaican, warm and connected, loving and curious. As they say in Jamaica, she hugged me up. And she loved the watch.

I turned thirteen that summer, and for the first time I felt as if I came from a real family. I met Harrisons; Aunt Cutie's son Calvert; Uncle Peter, brother to Grandma Sarah, and many of his ten kids; more cousins; the Tavares brothers, who owned a fabulous jewelry store in downtown Kingston; and quite a few others.

But I didn't meet anyone named Lowe.

THE YEAR
OF MY DISCONTENT

MY BROTHER HOWARD WAS UNINTERESTED IN HIGH SCHOOL AND AT THE AGE of sixteen contemplated dropping out, which he eventually did. My brother Elrick went to Williams College when I was twelve years old. College was something new in our family and all I knew about it was that, like high school, it lasted for four years and that you went away to it. Logically, then, I assumed, when Elrick was packing to go to Massachusetts, that he would be gone for the next four years. I was heartbroken. I cried into my pillow, disconsolate, throughout the night. How could I possibly survive for four *whole* years without my big brother Elrick? Today, Elrick will never let me forget that, after laughing at my foolishness, he had to reassure me that he would be back in a few months for Thanksgiving and Christmas.

The family was growing up and changing, and I was heading off to Cardinal Spellman High School, the most prestigious Catholic high school in the city of New York. The school was archdiocesan and its students came from the two boroughs that were included in the archdiocese: Manhattan and the Bronx. There were about two thousand students at Cardinal Spellman, and it was co-institutional: the boys occupied one half of the building and the girls the other.

There were no more than about eighty Black students at the time.

I was baptized Roman Catholic because of a deal my parents had made. She was Anglican and he was Catholic. In order for them to marry, she had to commit to raising their children Catholic. I had wanted to attend public high school, like my brothers. They knew better and worked to stop that plan. I argued but to no avail. I was furious, but there was no changing their minds. Then, like someone entering what Elisabeth Kübler-Ross has called the bargaining stage, I struck a deal with Nell and her sons. The entrance requirements for New York City's Catholic high schools involved a series of tests called the "co-op exams" that lasted an entire morning. I told them that if I was admitted to the most exclusive, highest-ranked Catholic high school, I'd go. Otherwise, a public high school—George Washington, Brandeis, or Julia Richman—would be welcoming me to its freshman class. I took the exams and aced them: my nature and character would not have allowed me to deliberately give incorrect answers. I was admitted to Cardinal Spellman High School and had to agree that Nell and my brothers were right. It set me on the path to success.

I entered Cardinal Spellman in 1966. After a few months, I was comfortable enough in my homeroom to sing (badly!) the hits of Diana Ross and the Supremes, Martha and the Vandellas, and other Motown girl groups. My classmates loved my versions of the accompanying choreography and frequently joined in (badly!). I was accepted not just as a fellow student but sometimes also as a real friend. That made my freshman year at Spellman fun, challenging, and welcoming.

By my sophomore year, the Black Power movement had begun. We joined the Students Afro-American Society (SAS), and this after-school club served as a place where the girls at Cardinal Spellman could explore our Blackness not only with one another but also with the boys. I was elected its president and led discussions about slavery, violations of civil rights, racist prisons, the corrupt justice system, and the organizations of Black people who had vowed to change all this, "fight the power," and even "off the pigs!"

During the summer between my freshman and sophomore years, the Paula Williams who'd shimmied and sung to the Supremes' "Stop! In the Name of Love" had been left behind—or, more accurately, she evolved into a Paula Williams who was becoming politicized, racially aware, and conscious of being in the vanguard of the fight for racial justice for African Americans. As I look back now, I can see how my fellow students and my teachers were confused by this transformation. But I was very clear about it. The time for foolishness was over: Blacks were being beaten, accused, jailed, and shipped off to an unjust war in Vietnam, and I no longer had time for singing and dancing.

It was a difficult year of high school for me and in some ways the most militant year of my life. Also, it was not a particularly happy time. I was devouring books like *One Hundred Years of Negro Lynching* and *Up from Slavery*.

Historically, 1967 was not a watershed year like 1968, but it was hugely important in the civil rights movement, and I felt the nation's tension within me. In April, Stokely Carmichael, a leader of the Student Nonviolent Coordinating Committee (SNCC), spoke at a big meeting in Seattle and used (possibly coined) the seismic term "Black Power." The shock waves were felt across the country because he was explicit about what Black Power meant: not only was it an assertion of black pride; it also meant "the coming together of Black people to fight for their liberation by any means necessary."

The phrase "by any means necessary" especially rattled the establishment.

That summer, the Supreme Court declared it unconstitutional to ban interracial marriages; such prohibitions were then still in force in sixteen states. But the summer was not a peaceful one. In Detroit and in Newark, New Jersey—New York City's neighbor—race riots exploded. The stakes, it seemed to me, had never been higher.

By this time our family had moved to an apartment on 110th Street between Manhattan and Columbus Avenues, and my best friend in high school, Sherry Bellamy, and I started visiting the nearby offices of the Black Panther Party in Harlem, on 112th Street and Seventh Avenue. Sherry's older sister was a card-carrying member of the Black Panthers and her father had been in the Tuskegee Airmen, so her family represented important extremes in the African American experience. Sherry and I had dismissed the "turn the other cheek" philosophy in favor of Carmichael's "by any means necessary." I had crossed an intellectual and an emotional line. Important things were going on in our country and the implications affected me and my family directly. These were battles of liberation and I truly suffered as I fought them.

My mother saw and understood the change in me. She appreciated how oppressed our people had been and encouraged me each time I spoke out, each time I confronted a racist position or person. It might seem surprising that my mother, who could otherwise be seen as strictly rule-abiding, had a radical soul. She joined me in my political awakening and radicalization. Though she looked Chinese, she identified herself as we her children identified ourselves: first "colored," then "Negro," then "Black." This identification included a fundamental sense of oppression and an unyielding need for rebellion. She supported me.

Some evenings my mother and I sat and watched as, on television, the civil rights movement unfolded: the water hoses and the dogs, the peaceful protests and the violence with which the protesters were met. There, in black and white, was the most important drama of our generation, as millions of men and women and children finally rose up to say "Enough." The great figures of the period—the Reverend Dr. Martin Luther King, Malcolm X, and Stokely Carmichael—inspired profoundly different reactions in my Hakka mother. Certainly, she appreciated that a social revolution has many phases and comes from many directions—she had seen this in Jamaica, as it wrested power from its colonial rulers—but she could not quite grasp the concept of "passive resistance."

Passive resistance? Were they crazy? In my knife-wielding mother's world there was never anything to be gained from being passive. Malcolm X made sense. Martin Luther King did not.

Something else did not make sense either. I returned home one day from the barbershop at the Harlem YMCA with my long hair shorn into a short Afro. My brothers joined in my mother's shock and disapprobation. She began screaming at me, but her invective was drowned out by the bass voices of both my brothers, jeering and enraged: *"What the hell did you do to your hair?"* They were so outraged that they wrestled me to the ground, not quite playfully, and punched me. The ostensible teasing did not camouflage their utter disapproval and disgust. Once more my hair was an issue in our family, but this time I wasn't a small child sent to another woman's house for its care. This time I was a proud, defiant adolescent. I knew that I was leading the way—and in fact, several years later my mother too cut her hair short; it curled around her Jamaican-Chinese face.

Like many members of my generation, I was breaking out in 1967, and the experience was disruptive, both disorienting and reorienting. I was a sophomore in a predominantly white Catholic high school, and although I could feel the power of the civil rights movement all around me, its heat did not penetrate the walls of my school except when I raised the temperature. I began to experience visceral rage over the oppression of my people. When I thought of British colonial rule in Jamaica—an experience far more immediate to my family than the American experience of slavery and the Civil War— it enraged me even more.

One day in a history class we were discussing current events. Mayor John Lindsay was facing one of New York's recurrent budget crises and had deemed it wise to shorten library hours by way of economizing. A fellow student suggested that curtailing library hours would have the unpleasant effect of a shared sacrifice, and it would be much smarter to "go to Harlem and get money from those welfare cheats."

I raised my eyebrows. I could feel my two Paulas begin to split apart, but I decided to let the conversation expand and continue.

One student after another agreed, pointing out the criminal nature of welfare and its recipients. Every possible stereotype of those in need of public assistance was raised: lazy, dishonest, living in idle luxury, possessing grand consumer durables purchased on the backs of good, honest, taxpaying citizens. I glanced at the clock. Class was going to be dismissed in a few minutes. If I was going to say something, I needed to say it now.

I raised my hand. The history teacher, Sister William Mary, called on me.

"Do you know anyone on welfare?" I asked.

The conversation stopped.

"Well, *I* am on welfare," I said. "I have been on welfare since I was five years old."

This was not the first time I ever made white folks uncomfortable, but it was one of the most satisfying. I could feel my energy rise.

"Have you ever *in your whole life* gone to bed hungry? Have you ever had social workers go into your house looking for a man because they thought your mother was hiding her lover?"

The silence was deafening. Needless to say, no one raised her hand.

"We never bought a Christmas tree until I was *thirteen* years old," I said. I didn't tell them that now we could buy one because my brother Elrick, who was on a scholarship at Williams College, sent most of his stipend home and we could finally afford to "waste" money on a disposable tree.

"We never had enough money. We waited until Christmas Eve, late, and then got whatever was left behind on the Christmas tree lot."

The class remained stunned into silence. And then the tears and whimpering began: "We didn't know that you're on welfare." "You're one of us." "I'm so sorry; I didn't know." I was both icy and fiery, and I was just getting rolling.

"How can you say some of the prejudiced and hateful things I just listened to?" I asked. "You don't even know the people you are condemning."

The bell rang. The clock showed that class was over. No one moved. I gathered my things to leave. As I walked down the aisle I said, "By the way, I am *still* on welfare."

Sister William Mary asked me to wait a moment. "Paula, that was so impressive. The girls needed to hear that," she said. "Thank you. In fact, I have another class this afternoon. Do you think that you could come and speak there as well?"

I suppose that she meant well, but I could hardly believe what I was hearing. "Excuse me, Sister," I said. "I have another class this afternoon. And God did not put me on this planet to teach white people what it is like to be Black and poor." There was another important point to be made: I asked her, "And when I'm sensitizing your students this afternoon, who is going to take my biology class that I would be missing?" She reddened. "No, Sister, I won't," I said, walking away. "And by the way, thanks for not speaking up during that discussion and not correcting or challenging their hateful statements."

Maybe I felt something else during that time, some unsettling telepathy from halfway around the world. On May 13, 1967, when he was seventy-eight years old, my grandfather died in China.

PART FOUR

SEARCHING FOR SAMUEL LOWE

To forget one's ancestors is to be a brook
without a source, a tree without a root.

—CHINESE PROVERB

THE YEAR OF THE DRAGON

I N THE CHINESE CALENDAR, EACH YEAR IS ASSOCIATED WITH A PARTICULAR animal; 1984 was the Year of the Rat, for example, and 2013 was the Year of the Snake. Rather than divide the signs of the zodiac into monthlong periods, as Western astrology does, so that each month is a different sign, the twelve animals in Chinese astrology each represent a single year; thus the cycle restarts every twelve years. The Chinese New Year sometimes occurs in January and at other times comes in February; the year rarely begins on January 1. If you are born, say, in January or February, your animal could be from the previous year. All this gets even more complicated with the addition of elements, colors, and other characteristics that are essential to the full astrological profile. But basically, each animal represents certain personality characteristics. Sheep, for example, are intelligent, artistic, and cultured, but they can also be insecure and self-indulgent; the rabbit is diplomatic, easygoing, and refined, but can also be detached and aloof.

I was born in 1952, which was the Year of the Dragon. The dragon is the only mythical animal in the Chinese calendar; the other eleven are real. The dragon's mythological status may help explain why it is considered the most powerful of all the astrological signs, and why dragons are very intense, strong personalities. Those of us born in the Year of the Dragon are self-confident, natural leaders, and we tend to be blessed with good luck and great prosperity. No argument here—but our negative trait is a tendency to be quick-tempered, arrogant, and obstinate. As it turns out, I am not just any dragon. The coordinates of my birth—the year, month, day, and time—come together to make me a *green* dragon, and this heightens the intensity of an already intense sign.

I have lived much of my destiny as a green dragon throughout my academic and professional career. I may have been pushing the limits at Cardinal Spellman, but I never slacked off, and I chose Vassar College from among several prestigious schools that offered me a spot. I was a dual major in history and Africana studies and planned a career in teaching. In fact, when I was a senior, I was accepted at Columbia Teachers College. One of my friends at Vassar was an African American woman named LaFleur Paysour, a couple of years older than I. She had gone to Columbia, but to the Journalism School. When she returned to Vassar for a visit, she said to me, "You should really think of going into journalism." This was during the golden post-Watergate period, when journalism had become a prestigious profession, one that attracted the best and the brightest. It was also one that was not exactly overpopulated by African American women.

"What's that?" I said. I was not being disingenuous—I really had no idea how newspaper stories were created or what it actually meant to be a reporter.

LaFleur laughed and explained to me the career that would eventually become my life. It sounded thrilling, commensurate with my energy and my curiosity. This was work where it would be part of my job to keep learning; I would not be frozen in teaching the same subjects for years on end. That conversation changed my life. I went to journalism school in Syracuse, met and married my daughter's biological father, and became a reporter at the *Syracuse Herald Journal*. I loved it. I loved talking to people, investigating injustice, looking at documents, and making connections. I was then hired by a larger newspaper in Fort Worth and made the transition into television eight years later.

Then the green dragon's need to organize and lead found its greatest expression. I worked at a number of television stations, always making my way up the ladder, until I moved to New York in 1989 with Roosevelt and my daughter Imani. I became first the assistant news director and then vice president and news director at WNBC. We took all of our newscasts to number one for the first time in sixteen years, and we won every television news award. In 2000 I was made the president and general manager of KNBC in Los Angeles, a city that was in many ways even more challenging and more diverse than New York. It was also a city whose proximity to China gave it a Chinese community that felt different from the one in New York. The Los Angeles community was more connected to China—closer to China—in many ways.

Whether by chance, coincidence, or destiny, 2012 was the most momentous year of my life. I had retired from NBC Universal the year before, and I was just getting used to my freedom. When 2011, the Year of the Rabbit, became 2012, became the Year of the Dragon, I could almost feel a quickening in whatever I did.

It began on a November afternoon in 2011, when I sat in front of the computer in our Los Angeles home and stared at the screen with the same intensity I have often seen on the faces investigative reporters. It is an expression that can be roughly translated as, "How the hell should I begin, because I *must* get the story?" It was raining outside. My husband, Roosevelt, had put on some music that I wasn't paying attention to, because I was preoccupied by the noise in my head. I wanted to find out what happened to my grandfather Samuel Lowe. I was no longer in a position to delay. I no longer had the excuse of a very busy life, because I had retired. I told everyone who asked about my plans that among my many projects—managing the family interest in our WNBA team, the LA Sparks; managing our interest in the

Africa Channel; and so forth—finding out more about my mother's family, and trying to discover the fate of her father, was my priority.

There I sat, poised to embark on this search. It was a strange, not very comfortable experience, suspended between knowing what I needed to do and actually doing it. I had often talked about finding my mother's family, but until very recently I'd had professional responsibilities that made it impossible to search in earnest. But that moment in November had a different quality: more momentous, less forgiving.

I tapped out, "S-A-M-U-E-L L-O-W-E-, J-A-M-A-I-C-A."

There appeared the South Carolina Historical Society's *Genealogical Magazine*, which mentioned that on "August 26, 1863, Samuel Lowe and John Harris of Port Royal, Jamaica, executed a power of attorney to Capt. John Flavell to collect all debts due." If only Samuel Lowe were a less common name. August 26, 1863, was over twenty-five years before my grandfather was born. More entries—less and less helpful—appeared. The World Wide Web does not get you far if you lack a distinctive name, and significant identifiers, of the person you are looking for. I needed to start closer to home.

I called the few aunts and uncles I knew, all relatives of my father. His brother and sister—Uncle Harry and my aunt Carmen, who is known by everyone as Aunt Cutie—are (respectively) sixteen and nineteen years younger than my father. "I want to find my mother's Chinese family," I said when Aunt Cutie answered the phone. I learned when I was a reporter that it is always a good idea to start an interview with a question that you know can be answered easily. The first easy answer seems to unseal other memories, perhaps because it sets off synapses—recollections—that might otherwise not fire.

"Maybe, to start, you could tell me exactly where Grandma was born." I meant my father's mother, Sarah. If I could begin the search with Sarah, I thought, then Sarah's relationship with my mother might offer some new information. Maybe Sarah had told her younger children stories about Nell that I had never heard. Cutie and Harry are elderly and wonderful, full of love and touchingly eager to help. But they were not able to remember much that was useful to me. Or maybe I just wanted more than anyone could possibly deliver. "My dear, I am not sure that I can help," Aunt Cutie said. "But Cousin JJ who lives in Toronto might have some answers. He knows a lot of the Lloyd family's history and I'll give him a call."

I hadn't talked to JJ in decades, but I knew that this didn't matter at all. Now in his seventies, JJ is John Hall, and he was my father's first cousin and Charlie Meade's younger brother. He attended Wolmer's, a prestigious preparatory school in Kingston, Saint Andrew's Parish, that was founded in 1729. Wolmer's educated and groomed many of the elite in Jamaica, and its motto—*Age quod agis*," "Whatever you do, do it well"—has inspired generations of Black Jamaican and Chinese Jamaican students.

Toronto is home to a large Jamaican and Chinese Jamaican community, so Wolmer's often holds alumni reunions and other events there even though it is a thousand miles to the north of Jamaica. After Aunt Cutie told him about my search, JJ sprang into action while attending one such reunion in Toronto in February 2012. He was a man with a mission. He worked the room, saying to those who were there, "I'm trying to help my cousin find her Chinese grandfather's people. Last name was Lowe. Do you know any Lowes?"

Surprisingly, one man responded, almost nonchalantly, "Oh, sure, but none of them are here."

The Lowes, he explained, like the Chinese gathered at the reunion, were part of a Chinese minority called the Hakka, an itinerant, culturally distinct group originally from northern China who were driven south by war and famine millennia ago and who were famous for their entrepreneurship and willingness to move to other regions in search of prosperity. Another distinguishing trait of the Hakka was that the women did not bind their feet as other women in China did. Hakka women were born entrepreneurs, practical, and decidedly *not* interested in showing their wealth by binding their feet and having to be carried or wheeled. This alumnus of Wolmer's also said to JJ, "Tell your cousin to come to the conference. There will be Lowes there." That was a stroke of good luck: JJ called me to say that the Hakka Chinese conference was planned for June 2012 in Toronto and he was sure I'd get many important leads there.

Around the same time, JJ met Carol Wong—an elegant, efficient woman whose father ran one of the most prosperous "Chiney shops" on the island. Carol grew up as part of this enterprise, which eventually included a complete dry goods store and a gas station. She had become the main point of reference for most of the Hakka community in Toronto. I e-mailed her, and we arranged to talk. I was as nervous as a rookie reporter when I picked up the phone to call her, but her voice was warm, friendly, and almost preternaturally calm. I told about my mother, including all the details I could summon, which clearly were not many. She listened quietly and then said words that I had spent my life longing to hear: "Paula, we will help you find your people."

THE HAKKA PEOPLE

THE HAKKA PEOPLE ARE EVERYWHERE. THEY TRAVELED FROM THEIR HOME IN southern China to the Caribbean, to the United States, to Canada. I have even heard that there are Hakka in Lesotho and in South Africa. Hakka means "guest people" in Chinese, and that translation is not a mere metaphor, a nice turn of phrase, but a fact. They are a guest people, nomadic, resilient, and distinct.

Scholars say that they originated in Jiangxi and began moving south from central China as early as the fourth century after Christ. They seem never to have stopped migrating for the next fourteen hundred years. In the eighteenth century some drifted into Guangdong and southwest to Guangxi, where they clashed with the local people over land and employment—this sounds rather like the Jamaican experience. Tension in the region rose; conflicts erupted; the Hakka continued to be marginalized and to marginalize themselves. Those who settled in the nineteenth century in Sze Yup, a region in the Canton (Guangzhou) delta, also created tension, and the region became overpopulated. A fourteen-year war ensued, from 1854 to 1867, between the Hakka and the earlier settlers, and hundreds of thousands of people died.

The scale of death in China over the generations is almost

unimaginable: hundreds of thousands of people here, a few million there. But even in this blood-drenched country, there were few events more devastating than the nineteenth-century Taiping Rebellion, a civil war that lasted from 1850 to 1864. The ruling Qing—or Manchu—dynasty had been marked by defeats in battle, economic hardship, various episodes of social unrest, and the Opium Wars. The leader of the rebellion against the Qing, a charismatic Hakka named Hong Xiuquan, was perhaps not the proudest exemplar of the Hakka people. He was literally messianic, believing he had learned in a vision that he was the younger brother of Jesus; and by 1848 he had attracted ten thousand followers, mostly from the Hakka and other minorities in the region. By 1850 the number of his followers had grown to more than twenty thousand. This army defeated the government troops, and in 1851 he declared himself Heavenly King of the Taiping Tianguo. The rebellion lasted another thirteen years and claimed twenty million lives. Nowadays some scholars put the estimate at forty million.

Among Hong Xiuquan's recruits were Hakka women, who actually fought in the battles. When his rebel troops occupied Nanjing, these Hakka women, with their weapons and their enormous unbound feet, were a shock to the more traditional army.

Strong independent women.

Fierce fighters.

Unbound feet.

Of course my mother was not just Chinese. She was Hakka.

CHEE GAH NGIN:
MANY PLACES, ONE PEOPLE

THERE WERE SEVERAL LOWES IN TORONTO WHO WOULD ATTEND THE CONFERence and they were all Hakka Chinese. In fact, one of the cochairs of the conference was named Keith Lowe, Carol said. Could he be a blood relative? Could he know what happened to Samuel?

I typed out the registration forms for the conference for me and my brothers. The conference was being held at York University in Toronto from Friday, June 29, to Sunday, July 1. This was good timing, because a friend and I had planned to visit China on August 9.

JJ had also described a thirty-minute documentary called *The Chiney Shop* that he had seen at the Chinese Cultural Center of Greater Toronto, and he was sending me a copy. I couldn't wait for it to arrive, and as soon as it did, I put the DVD into our player. It was about Chinese merchants in Jamaica—their invaluable memories, the tension and conflicts they endured, the money they made, the good they did, the lives they led. These were my mother's and grandfather's people. The documentarian, Jeanette Kong, a Canadian journalist and filmmaker, had grown up in her parents' Chiney shop, like Carol Wong and like my mother.

I knew that if I spoke to no one else, I needed to meet Jeanette

Kong. I had never glimpsed the world into which my mother had been born. I knew that her father was a merchant, but now I understood the larger context of that occupation and how it had shaped Jamaica in the nineteenth and early twentieth centuries.

Shortly after I received the video, my father's sister Ouida called from Jamaica.

"Paula, there's a woman who walks past my house for exercise, and her name is Lowe," she said. I began to feel as if I were on an exciting ride in an amusement park, as if I were being carried and caromed in all directions. After so many years, how could all these forces start coming together at the same time? The power of the Year of the Dragon was taking hold.

"She's Chinese," Aunt Ouida continued. "I am going to stop Eily Lowe one day and ask her if she might be your mother's relative."

Lowe turned out to be this woman's married name, but even so she was helpful: she told my aunt that her husband's cousin Raymond Lodenquai lived in Toronto and was steeped in family history. She gave my aunt his e-mail address and I contacted him. He responded immediately and explained that his family, and many other Lowes, came from a place called Niu Fu. Unfortunately, this year he couldn't attend the Hakka conference. Still, he had already set me thinking: Could my grandfather's Chinese name be Lodenquai? And did Samuel return to Niu Fu?

The theme of the conference was "Many Places, One People." On Friday night there was a reception at the Chinese Cultural Center. On Saturday, attendance had swelled to about four hundred and there was a keynote lecture, "Legacy and Continuity." There was also a presentation of a History Channel documentary on Hakka *tulou*, unique pounded-earth buildings in which a circular fortification encloses several residences. Breakout sessions focused on the Hakka diaspora in Taiwan, Malaysia, Calcutta, and Hong Kong.

My brothers and I were a small though conspicuous knot of African Americans—with a television crew—in a sea of hundreds of Hakka Chinese people. As the other participants learned why we were there, they responded in ways I had never seen before: no eyebrows were raised; no one looked incredulous. But the conversations took a familiar turn when they asked for my grandfather's Chinese or Hakka name.

"I don't know his Chinese name," I said, as so often before, thinking, *God, if I only knew his Chinese name.*

They asked, "Well, where in Kingston did he live?"

"I don't know."

"Where was his shop?"

"I don't know."

And then came something beautifully new.

"OK, OK, don't worry," they would say. "We're going to help you find him." They were not disheartened or discouraged. For the first time in my life, neither was I.

They scanned my mother's photo as if she were a missing person—in a way, I suppose, she was—and sent it out across the Hakka community in Toronto and Jamaica. Maybe someone would recognize her.

On Saturday morning my brothers and I stood in the conference hall with the documentarian Jeanette Kong, who made *The Chiney Shop* and would become an essential participant and my dearest confederate in our search. I told her that she had to help me find my grandfather and his people. "Look," I said, "I was not born in Jamaica. I didn't go to school in Jamaica. The Hakka community does not know my mother. Looking the way I look, I cannot walk up to Hakka people and say, 'Can you help me find my grandfather?' It's not going to work." But Jeanette is Hakka; she was born in Jamaica, grew up in a Chiney shop, and lives in Toronto. She knew all these people. "You have to help me," I repeated.

Immediately, she agreed.

She then began speaking a language that I soon realized was Hakka. She said some words as if she were talking to a child and suddenly my brother Howard chimed in with the same words. I whirled around to be sure that these sounds were actually coming out of his mouth.

"What's that?" I asked both of them. "What are you saying?"

Jeanette said that she was counting from one to ten. "How do *you* know how to do that?" I asked my brother. Howard looked both proud and embarrassed. "Ma taught me that when I was little," he said. "I didn't even know that I knew it. It just came back to me." But now an image came to me: my two- or three-year-old mother, sitting with her father as he patiently taught her to count to ten. This was a lesson that she never forgot, a gift from him, perhaps more valuable than the pearl earrings he was said to have left for her. Indeed, it is a gift that was notable for its very ordinariness as part of a timeless, universal ritual between parents and children.

Nell's experience of her father was pervaded by loss, but her fragmentary knowledge of Hakka was a genuine connection to him. And for me, it was as if I had found a letter he had written to her. Hearing these numbers in a mysterious language, I turned to Howard with newfound respect. He was handsome, dark like our father, and there he was, counting in the language of our Chinese ancestors.

The next day as my brothers and I were saying our farewells to a small group with whom we had grown close, one of the Hakka community leaders, Patrick Lee, asked if we knew when our grandfather had arrived in Jamaica, or in what year he had returned to China. "Not a clue," I responded, certain of my facts (or lack thereof).

At the same time my brother Elrick responded, "1934."

Another revelation! "How do you *know* that?" I asked.

"Ma told me," Elrick replied. Actually, she had told him that she was almost sixteen when her father left, and Elrick did the math.

"Well, if your grandfather left in 1934, he would have been quarantined at Ellis Island," Patrick Lee observed.

"No. No one in my family went to the United States until my

mother arrived in 1945 under the Chinese Immigration Act quota," I said decisively.

Here I felt I was on firmer ground. I had received the Ellis Island Medal of Honor, with about thirty other "outstanding children of immigrants," when the impressive registry of immigration was first opened. The officials had invited several luminaries—children and grandchildren of immigrants—to the ceremony, and I had watched as Madeleine Albright, Congressman Charlie Rangel of Harlem, and Hillary Clinton searched through the online computerized archive to find their relatives. I had felt awkward and orphaned, knowing that no relative of mine was in the registry, that there would be no trace of my grandfather in the archive. While others were exclaiming excitedly as they discovered parents or grandparents, I stood by a window, struggling to contain my tears, and failing.

Patrick, however, knew more. "All aliens traveling through U.S. waters had to be quarantined for a few days at Ellis Island," he said. "Go online and look at the immigration records and you are likely to find your grandfather's name."

Although the conference was not over, my brothers and I were departing for a flight to Los Angeles and Elrick's drive back to Chicago. As Howard and I sat in the lounge at the Toronto airport, I thought about how things were coming together. My mother's photograph was out in the Hakka community. I knew Jeanette would be able to find some connections. It might be slow and laborious, but for the first time I had confidence that we were moving toward some answers. At the conference I had learned that the Chinese who went to Jamaica mostly originated in Guangdong Province. Now, in the airport, I accessed the National Registry from Ellis Island on my iPad and typed into the search function, "Samuel Lowe, Guangdong, China, Kingston, Jamaica, 1934."

A ship's registry appeared on my screen.

His name was in the registry, but for 1933.

I was calm but felt rather breathless. Howard turned to me, sensing something. "What's wrong?" he asked. I whispered, "I think this is him."

"Who?"

"Grandpa. I think that this is him," I said, pointing to the name "Samuel Lowe" on the screen. My hand was shaking.

THE CAPITALIST CLASS
AND THE SERVANT CLASS

EARNING ABOUT RACE AND CLASS IN JAMAICA INVOLVED *UNLEARNING* MUCH of what I knew about race and class from my American perspective. In my search to understand the lives of my grandfather and my mother, I had many revelations about how Jamaican society functioned, what impact the Chinese had on that society, and why the little Chiney shops in each community were so important economically and socially—they were microcosms of a larger world.

After years of indentured servitude, many Chinese eventually became merchants and leading retailers. The British government had placed no restrictions on the movement of the Chinese immigrants, and so these disciplined entrepreneurs scattered throughout Jamaica, from tiny villages to big cities like Kingston. In almost every community, there was a Chiney shop, where most of the residents bought groceries, dry goods, and odds and ends. The shops had a distinct look: a large overhanging roof to protect the customers and proprietors from the rain, two large shuttered windows, the shop itself in the front of the building, and the family home in the back.

The shops sold staples such as flour, cornmeal, grains, and rice; various kinds of breads; and canned goods. Some shops—I believe my grandfather's was one of them—also sold pots and pans, fabrics, needles and thread, hammers and nails. Simple clothes would be hanging in racks. Sometimes, enormous glass jars full of sweets stood like sentinels by the cash register. Coconut cakes, tamarind balls, animal crackers, and barrels of salted fish tempted customers as they walked down the aisles.

Carol Wong, who now lives in Toronto, grew up during the 1950s in one of the island's most prosperous Chiney shops. She recalls, in Jeanette Kong's documentary *The Chiney Shop*, that her father began, like my grandfather, with a simple dry goods and grocery store. "Eventually we went into hardware and expanded into a rum shop," she said. "The locals loved the rum. Then we added music, with a jukebox room, so things got a bit rowdier. When cars came to Jamaica, we added on to this concrete block building. First we had a gas station. Then we added on a theater." I could only wonder what my grandfather might have created, had his fortunes not turned.

Credit was crucial for those who lived and worked there, and relationships developed between the Black population and the Chinese population. Buying and trading, stopping to chat, looking for one item or another—all this eventually led to more intimate relationships, like Samuel Lowe's with Emma Allison and Albertha Campbell.

The intimate relations typically involved a Chinese man and a Jamaican woman, because Chinese men were overwhelmingly the ones who traveled to Jamaica. They were brought in as laborers and stayed to take advantage of economic opportunities. Jamaican schoolchildren often sat next to a half-Chinese classmate. There they would sit, the shopkeeper's child alongside children from the hills. In Jamaica it was simply not strange for a Chinese and a Black person to be courting, or planning to get married, or having children.

I have heard from many Jamaican friends that a beautiful trait of the Chinese who came to Jamaica was their adaptability. The Chinese mixed easily into the culture and began to speak the Jamaican patois with a particular Chinese lilt. For many Jamaican children, going to a Chiney shop after school would mean receiving a little candy or some other free treat. The generosity of the shopkeeper was also an incentive for children sent on errands by their mother to shop at the Chinese store.

But there was always a dichotomy. Some Jamaicans loved the Chinese; others despised them. I suppose it's a tragedy of human nature that we tend to regard with suspicion those who don't resemble us. In the context of colonialism, suspicion can intensify into hostility. The age-old pattern of oppression—those in power demeaning the less powerful, and the less powerful in turn demeaning the least powerful—was enacted in Jamaica. The British colonialists maligned the African-Jamaicans, who then aimed their own contempt at an obvious target, the Chinese population. The Chinese were so distinctly different, in their customs and appearance, that strange rumors circulated, and children imagined that the Chinese standing behind the counter had no feet. Even today, tales are told that the Chinese eat dogs in Jamaica.

And yet the Chinese seemed to appreciate the implications of many Jamaicans' poverty. At that time, the rhythm of daily life in the small towns and even in Kingston did not involve planning ahead or buying for the week. In the morning, women might go to the store, or send a child, to buy something for the family breakfast. In the afternoon, another child might appear with the dinner list. The proprietor of a Chiney shop would have a small accounts book listing each client's name and balance. On the many occasions when money was short, customers would buy what they needed and sign for it until the next paycheck, when the debts were, if not cleared, then lessened.

It was, overall, a mutually beneficial, smoothly functioning commercial system. When someone couldn't pay and a debt was not honored, that became a cost of doing business. But bad debts were exceptional. The system worked because everyone played his part. And in consequence, trust and even friendship blossomed between the Chinese and the Blacks in Jamaica.

Still, there were times of conflict and discontent, when tension could not be ignored and violence resulted. Not all the Chinese were benevolent shopkeepers. Some competed for jobs and opportunities; some ran their businesses with less integrity and exploited their clients, especially in 1918 and sporadically in the 1920s and 1930s.

I found a report from August 1918 by a police inspector in Saint Catherine parish. He described vividly how simmering social issues could become explosive. This incident began when a police inspector "visited the room of a girl servant employed to a Chinese shopkeeper there." (I read this as sexual assault by a policeman, but that was not explicit in the report.) The policeman was discovered by the shopkeeper "and apparently, stupidly did not divulge his identity to the Chinaman—to whom he was well known—being in plain clothes at the time." The Chinese proprietor began to beat the policeman, and was joined by a few others. The policeman clearly knew he was in the wrong, and went into hiding for two days. "His disappearance caused some excitement in the district and the lower orders made out that he had been killed by the Chinaman. Giving this as an excuse, big crowds assembled in the village . . . and started to beat down all the Chinese shops there and loot the goods." The unrest spread from town to town as word circulated throughout the area, and when it was all over, in five days, dozens of shops owned by Chinese men had been destroyed. My grandfather's shop was not included in the list, but the riots must have frightened him and made it clear that life might not always be peaceful in his new home.

In July 2014, I revisited my grandfather's property in Mocho, Clarendon, Jamaica, along with seven of my Chinese first cousins and second cousins. They traveled from China, Hong Kong, Australia, the United Kingdom, the United States, and Jamaica to see the home that he gave to Emma Allison and where his children with her, Adassa and Gilbert, were born. This house, which is still standing and is home to Emma's granddaughter and family by a man who entered her life after Samuel left her bed, is situated on a hill adjacent to a police station. During that visit, we learned that Samuel donated the plot of land adjacent to his house in order for the town to build its police station.

Once again, I stumbled across a fact that proved to me how strategic and intelligent a man my grandfather was. His shop was across the street from his home *and* the police station. Unlike some of his fellow Chinese shopkeepers, Samuel Lowe never was troubled by anti-Chinese sentiment in Mocho, Clarendon, and his family and shop remained safe.

SAMUEL LOWE
WAS MY FATHER

"THIS IS YOUR GRANDFATHER," JEANETTE WROTE. "HE IS MARRIED AND HE HAD children. He was traveling with his wife and two daughters."

While I was flying from Toronto to Los Angeles, Jeanette got to work. She looked at the ship's registry I had discovered, and saw that just under the name Samuel Lowe were the names of his wife, Swee Yin Ho, and two of his children, with their ages. His two daughters were Barbara Hyacinth and Anita Maria, and there were also two sons of his cousins.

My mother had brothers and sisters.

My grandfather went to China but did not die.

These two facts came, not as a shock, but rather as a reassuring confirmation of what I had always believed. I *knew*, as surely as I knew my own name, that my mother's people were out there to be found. The confirmation opened another door, posed greater challenges, and raised more questions. These daughters, my mother's half sisters, probably had children. If so, there must be descendants somewhere. Where was this family in the vast expanse of China?

I learned from Hakka scholars that there were only two likely possibilities: my grandfather could have gone to Niu Fu or to Lowe Shui Hap. Everyone in Toronto told me that I needed to talk to a man named Winston Lowe, but he was ill and in the hospital. Another Lowe in Toronto, Keith Lowe, had been the cochair of the Hakka conference. We spent some time chatting about my quest for my family, and he later promised Jeanette that he would help me. The first step, he said, was to contact his nephew in China, who was also a Lowe and might have some ideas.

Around the same time that I discovered the ship's registry and Jeanette discovered the other children, Keith Lowe sent an e-mail to this nephew, Yiu Hung Law, and asked about Samuel Lowe, who had been a merchant in Jamaica. The next day, he received a response: "Uncle, let me ask one of my uncles if he's ever heard of Samuel Lowe. I will get back to you."

I expected that there would just be a few more connections, a few more helpful people, this time people who happened to live in China.

Two days later, still buoyed by what we had recently learned, I went to my computer. Nothing had prepared me for what I found waiting there; it looked like any other e-mail.

Yiu Hung had asked his uncle, Chow Woo Lowe, about Samuel. The reply in my in-box was:

"My uncle says Samuel Lowe was his father."

Yiu Hung's uncle says Samuel Lowe was his father.

My mother's younger brother is still alive.

I found him.

I found my uncle. And from my uncle, I will find Samuel Lowe.

And my Chinese family.

Was this really happening? Was the woman who resembled my mother—the woman on a street in Beijing during the Olympics— sent to me to lead the way?

DO YOU KNOW
THAT YOU ARE BLACK?

I **WAS EXHILARATED, DUMBFOUNDED, BEWILDERED, AND GRATIFIED WHEN I** returned to Los Angeles after the conference. I couldn't wait to share all that had happened with Roosevelt. He is an artist who came from a large, happy family in New Orleans. His life as he was growing up—devoted parents in a good old-fashioned marriage— was the polar opposite of ours. He had three brothers and a sister who helped one another, hung out together, had the predictable sibling dramas. We have been together now for almost thirty years. There is no one who provides the ballast in my life, the true north of my compass, the way Roosevelt does.

He had followed my quest with some detachment but also concern. I know that he wanted to protect me from a possible bitter disappointment, but at the same time, I could see that something was brewing for him. He was working on the problem.

On the morning after I returned home, my husband and I popped into the hot tub at four o'clock for some conversation and reconnecting. I was vibrating with excitement over the possibility that this journey would actually lead to a place of completion and fulfillment. I told him about the conference, about the people there, about being a part of the Hakka experience, about Jeanette and my discovery of the ship's registry. He sipped his coffee and listened.

"You know," he said. "This has now turned into an almost all-consuming passion."

"*Absolutely!*" I said. Perhaps I might have been a little nuts about it. I might have seemed to be on an emotional level from which I could only come crashing down. Roosevelt continued.

"Paula, what are you expecting to happen?" he asked.

"What do you mean?"

"When you find these Chinese people in your family," he continued. "What are you expecting will happen?"

I began to wonder where this was going. "I don't know," I said, making a slow descent back to earth. "What are you asking me? What do you mean?"

He gave me one of his serious, twenty-four-karat, all-Roosevelt looks. "Baby," he asked, "do you know you're Black?"

I looked at him, puzzled and defensive. "Yes. I know I am Black."

He looked a little hesitant, as if he were afraid to say any more, afraid to make the next point. He was worried about the unknowns. He worried that I might indeed find this family of mine, of Samuel Lowe's—find whoever might still exist. If I did manage to track them down, there was the unknown response: how they might react to me. There was a possibility, in fact a likelihood, that they would not share my enthusiasm for discovering long-lost relatives—long-lost African American relatives. They might see an African American woman and reject me. Until this moment, that thought had never occurred to me. I sat in our hot tub—with my ginger-brown skin, my proudly worn Afro—and, as my glasses fogged from the steam, wondered first why he was saying this to me and, second, shockingly, why I hadn't thought of it on my own.

I paused and inhaled slowly. "I expect that because I am their family, and they are my family, we will be family. That is all I expect."

He nodded gently, with a soft look in his eyes. And at that

moment, I realized his question came from his love, from his wanting to protect me, and from his experience as a man who had grown up in a racist, racially divided United States. I realized he was trying to prepare me for a possible disappointment, to soften a possible blow, to get me to bring down my intensity a few degrees.

He seemed to understand that time would provide answers to the questions that hovered in my mind. He was saying to me: You are passionate about this. You really want it. But in the end, it might not be what you want. It might not turn out to be what you are dreaming. And I knew he was trying to help me. To watch my back. To be the husband I have relied on for decades.

But in every fiber of my being, I knew several absolute truths:

That I didn't have to worry.

That I am a Lowe.

That I am African American and I am Chinese.

That the face of my mother is a Chinese face.

That her face was marked by kindness and protectiveness, pride and disappointment.

That she believed and lived by the maxim "Family above all."

That her family will want to know about my existence as much as I want to know about theirs.

PART FIVE

PROSPERITY, FAMILY, EDUCATION

A man of determination will surely succeed.

—CHINESE PROVERB

THE POINT
OF NO RETURN

IT'S AUGUST 2012 AND I AM IN MY ROOM AT THE SHENZHEN MARRIOTT HOTEL. I am trembling. Inside the temperature is seventy-three degrees and outside it's in the eighties, so obviously my shivering has nothing to do with the climate. My cousin Yiu Hung Law has just called my room: "Paula, where are you? I am here with my aunt and uncle waiting to meet you."

My dear friend Marcia Haynes and I had planned several trips to China, and they usually involved the glories of food, shopping, and music. This one was supposed to follow the same general plan, until I received the e-mail from my cousin saying that my uncle would be happy to meet with me. Then everything changed. Marcia and I knew that after we visited Beijing, we were going to Shenzhen.

Throughout the visit to Beijing, the thought that I would be seeing my family in three days, in two days, in one day, nudged me at odd moments: during breakfast, or when Marcia and I were looking at shoes, or just before I went to sleep. Or it might be the first thought I had upon waking. By the time we arrived at the gleaming Beijing airport, I was already nervous and excited.

After we checked in, I looked at the status board and discovered that our flight was going to be an hour late. I am compulsively punc-

tual, and even though delayed flights are obviously beyond my control, this one had particular implications. Not only would my own
wait for this meeting be extended by yet another hour—after all
these years, another hour?—but, more important, that my elderly
uncle and aunt would have to wait for me, and this was unacceptable.
I e-mailed Uncle Chow Woo's grandson Henry and told him that my
aunt and uncle should come to the hotel later, because I didn't want
them to have to wait in the lobby for me.

As I wrote those words, I was thinking: I am in China and, aside
from Marcia, who—like me—is Black, I am surrounded by people
who are Chinese. All around me were Chinese folks—businesspeo-
ple in a hurry; young mothers steering baby strollers; college kids in
the universal uniform of T-shirt and backpack; older couples carrying small satchels; sleek, beautifully groomed, modern young women
talking on cell phones—all of them making their way from Beijing to
other cities. They were oblivious of my physical presence—what did
they care about two Black female tourists? What struck me was my
heightened awareness of my connection with each one of them. Of
course, they had no idea that I was there to visit my family. Still, I felt
a surge of pride that I was communicating with people who were my
family and who looked like my fellow airline passengers. I was one of
them. I was transformed from a tourist to a local.

Henry, who was at work, responded immediately and agreed
to try to delay the entourage of family members who were bringing Aunt Adassa and Uncle Chow Woo to the hotel. Even though he
was not part of that entourage, I would be meeting him over the next
few days. The hour-long flight finally took off and our landing was
smooth. I didn't actually run to grab a cab, but I could feel my inner
excitement building. Fortunately, I thought, I would have time to collect myself once we were in the hotel.

The cabdriver pulled up to the Marriott, its lobby an amusing
hybrid of Chinese touches with the familiar Western amenities. I
checked in, took my bags to my room, and was going to spend a bit

of time freshening up, in order to make the best possible impression on my *family*.

I was not in my room for three minutes before the phone rang.

They were in the lobby waiting for me. They had called me! I had asked Henry to delay them, but that was impossible. They were as eager to meet me, it seemed, as I was to meet them. I have had many important meetings in my life, but absolutely none were as signifi-cant, as monumental, as life changing as this one.

I didn't feel ready.

I couldn't be late.

I called Marcia's room. "They're here," I said, my voice convey-ing all my anxiety, excitement, impatience, and happiness. She had never seen me, or heard me, quite so undone. "Look," she said. "I will meet you down there after I freshened up." (Is that what I was feeling too?) There was no time for any of it. No time to freshen up. No time to change my clothes. No time to grab all the things I wanted to take with me, such as the Sony camcorder I had bought for this occasion. I remembered to grab the file with my mother's documents, the leather five-by-seven-inch envelope that held photos of Nell over the years, my purse, and my room key.

I needed to grab my emotions as well. "Stop trembling," I ordered my hands.

All alone I descended in the elevator. Thirty-third floor . . . *Ma, they're here.* Twenty-ninth floor . . . *Grandpa, are you watching over me now?* Eighteenth floor . . . *They're here. Your children are about to meet your granddaughter.* Tenth floor . . . *Grandpa, Ma, are you hold-ing hands now? Grandpa, are you caressing Ma's face? I can feel you both smiling.* Fifth floor . . . *I can feel your joy. God, am I ever nervous.*

Lobby. Breathe.

I'm good. I stopped trembling.

The elevator doors opened and my feet had a mind of their own. Step. Step. Step. Step. Make a slight turn. Step. Step. A gray-haired man who was about my age approached me—clearly I was not hard to spot. "Paula?" he said smiling. "I am Yiu Hung and this is your uncle and this is your aunt."

I saw a graceful, much older man—with a fine-boned patrician face and a gentle smile—walking slowly toward me. Uncle Chow Woo opened his arms and even though I'm fully five inches taller than he, I felt enveloped and safe. Somehow my earlier worries and nervousness, my need to be prepared and impress them all, seemed ridiculous. It was happening and I was with my mother's—my virtually orphaned mother's—flesh and blood. We walked together to the lounge area, where the others were gathered and had taken over a carefully choreographed seating arrangement. A sofa and six chairs were arranged around a large chinoiserie coffee table.

My aunt Adassa sat in the center of the sofa. She was tiny, with my mother's touch of caramel coloring. Despite the warm weather, she wore a little woolen cap over her gray hair. With surprising ease for a ninety-three-year-old woman, she stood up and hugged me with an intensity and strength that I would not have imagined she was capable of. Nothing uncertain or restrained here—it was as if she were making sure that I was real and not an apparition.

Aunt Adassa had known she had a younger brother named Gilbert who had been left in Jamaica. For her, the notion that there might be family members she had never met was simply a part of her life. At the moment, she may have thought that I was one of Gilbert's children. But it didn't matter. She sat down again, beaming with happiness. Her youngest daughter and youngest son and her daughter-in-law, who had escorted her, sat to her left on the sofa and on its arm.

Uncle Chow Woo sat on Adassa's right, flanked by one of my cousins. I settled in the chair immediately to the right, perpendicular to the sofa and within arms' reach of Uncle Chow Woo. His son-in-law, his grandson-in-law, and Yiu Hung's son, Stanley, took the other seats. I looked around at the ten of us, at the range of colors, ages, and relationships, and I thought of the amazing history that had brought us all together. Tea had been ordered and was placed on the table.

Amsterdam Avenue, New York City, the early 1950s. Nell sits in our Harlem living room. Beautiful and brooding is how I most remember her.

145th Street, Harlem. Cousin George Barnes, my dad's first cousin, who sometimes rented a room in our small apartment, was the only relative we consistently saw, and he protected and loved us. My mom is on the right, holding the handkerchief. She likely made her beautiful dress.

Columbus Circle, New York City, circa 1966. Nell, Elrick, and me during the summer after his freshman year at college and as I was entering high school, on a rare trip to Midtown Manhattan with my mother. She mostly stayed in our apartment or on our tenement's stoop.

Harlem home of Conrad and Lotte Sweetland—112th Street and St. Nicolas Avenue. Howard, five, and me, three, visiting Elrick Jr.'s godparents. Our father was best friends with Connie and his wife, Nana, who cared for us as if we were their very own.

1958. We attended St. Rose of Lima School at 164th Street and St. Nicholas Avenue. My first-grade photo was taken when I had the mumps, and I was to remain at home for about a week afterward.

Me at age six with Howard, age eight, in our Harlem kitchen before heading to classes at St. Rose of Lima School. Elrick was our brilliant eldest brother. He set the stage for the Williams kids' being known for their intelligence—and for having a mother who was as protective as a lioness.

The birth certificate for my father, Elrick Mortimer Williams Sr. As was a common practice, the certificate lists an incorrect date of birth. This sometimes was done to "duppy-proof" a child—to protect him and make it difficult for evil spirits to find him.

Harlem, 1959. My father and me following my First Communion at St. Rose of Lima Roman Catholic Church.

Harlem, 1956. The only existing photo of our family, taken after Elrick Jr.'s confirmation at St. Rose of Lima Church. Nell was thirty-eight; I was four; Howard, six; Elrick Jr., nine; and Elrick Sr., thirty-nine. My parents' marriage was already a failure, and the looks on our faces showed the anxiety we felt for fear they'd quarrel, or worse, whenever they were in the same room.

China, circa 1929. A Lowe family portrait—Samuel Lowe (Lowe Ding Chow) and his wife, Ho Swee Yin, with sons Chow Woo, Chow Kong, and Chow Ying. We are now close to Uncles Chow Woo and Chow Kong, who greet us in Guangzhou whenever we visit. Uncle Chow Ying died in 2007. (*Courtesy of the Lowe/Luo Family Photo Collection*)

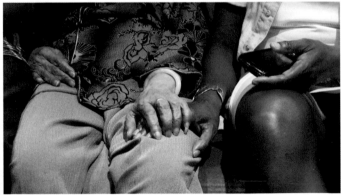

Shenzhen, August 2012. Aunt Adassa asks me to bring all of her nephews and nieces to China to meet the Lowe/Luo clan. (*Photo by Marcia Haynes*)

Guangzhou, December 2012. Aunt Adassa Lowe—the second eldest of Samuel Lowe's daughters—half Black Jamaican and half Chinese. Her skin coloring matches my mother's, the eldest daughter of a different mother. (*Photo by Elrick Williams*)

Guangzhou, December 2012. Aunt Barbara Lowe, the second youngest of Samuel Lowe's daughters. (*Photo by Elrick Williams*)

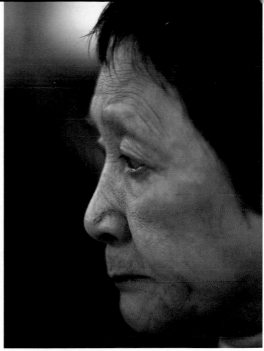

Guangzhou, December 2012. Uncle Chow Kong Lowe, the youngest of Samuel Lowe's sons. (*Photo by Elrick Williams*)

Lowe Swee Hap, December 2012. Yiu Hung Law (*left*) who asked the Lowe/Luo family in China if anyone knew of Samuel Lowe; Uncle Chow Woo (*right*) responded, "Samuel Lowe was my father." That e-mail ended my search for my family living in China. (*Photo by Elrick Williams*)

Shenzhen, China, December 2012. Uncle Chow Woo's granddaughters, Yuxin (Wendy) Wen and Siqi (Sharon) Luo and me. During that week, more than three hundred of my Chinese relatives came from Australia, the UK, the United States, Canada, Jamaica, and China to meet the twenty African-Jamaican-Chinese relatives descended from Lowe Ding Chow's long-lost children. (*Courtesy of the Lowe/Luo Family Photo Collection*)

Christmas Eve, 2012. My husband Roosevelt's fifty-sixth birthday party in Shenzhen with us surrounding the family patriarch, eighty-seven-year-old Uncle Chow Woo (center). I am at the far right, and Roosevelt is in the back row, third from the left. (*Photo by Elrick Williams*)

Guangzhou, December 2012. The reunion of my mother's siblings with some of their children and grandchildren, meeting the Afro-Chinese descendants of Nell Vera Lowe Williams and Gilbert Lowe. (*Courtesy of the Lowe/Luo Family Photo Collection*)

Mocho, Jamaica, March 2013. The Lowe/Luo family in front of Samuel Lowe's first shop, near a sugar plantation in Clarendon Parish. Descendants of Nell Vera Lowe, Gilbert Lowe, and Chow Woo Lowe are pictured here. Nell and Gilbert were the only two who did not go to China with their father. (*Photo by Elrick Williams*)

Guangzhou, April 2013. Samuel Lowe's eldest grandson, Elrick Williams Jr. (*far right*), is elected chairman of Lowe Family Enterprises. (*Courtesy of the Lowe/Luo Family Photo Collection*)

Guangzhou, April 2013. Cousins gather with Aunt Anita Maria (*seated, center*) to form Lowe Family Enterprises. I am seated next to her, and Elrick Williams Jr. is in the rear at the far right. (*Courtesy of the Lowe/Luo Family Photo Collection*)

Guangzhou, February 2014. During our Lunar New Year visit, Uncle Chow Woo was surrounded by some of his children as well as some of the children of Nell, Gilbert, and Barbara, and their first cousin Leslie Lowe.

My husband, Roosevelt Madison, who lovingly ended the awkwardness at the first family gathering at Lowe Swee Hap by ceremoniously toasting each table of relatives with Chinese sorghum liquor Moutai. It was a real party after that! (*Photo by Elrick Williams*)

Inflight to Shenzhen, December 2012. Imara Jones, Carlyn Jones, and Idris Morales—all great-grandchildren of Nell Vera Lowe Williams. (*Photo by Elrick Williams*)

My beautiful physician daughter, Dr. Imani Jehan Walker, and my grandson, Idris Morales, who inspire Roosevelt and me each and every day. (*Photo by Patricia Jordan*)

To make sure that we would be able to communicate, I had hired a Mandarin-speaking interpreter. I didn't know that my cousins Stanley had attended the University of Texas at Austin and Gary had attended the University of Massachusetts. In fact, Stanley studied English in elementary school, high school, and college in China. This was the new China. The young generation lived in a global context and was no longer isolated in a Communist society.

We exchanged pleasantries—they asked Marcia and me about our flight, I asked about their trip—but we were all eager to begin the process of finally getting to know one another, of solidifying connections that had been severed for nearly one hundred years. For my part, I had assembled a thick file of newspaper articles about my grandfather from Jamaica's *Daily Gleaner*, the birth certificates of my aunts and uncles who were born in Jamaica, a family tree that I had created at ancestry.com, and the passenger lists of the SS *Sixaola* from 1927 and the SS *Adrastus* from 1933.

Both ships sailed from Kingston to Hong Kong. The *Sixaola*'s list showed Samuel Lowe, thirty-eight, merchant; Swee Yin Ho, twenty-six; Adassa, eight; Chow (Ying) Lowe, four; Chow Woo Lowe, two; Chow Kong Lowe, one. This was the trip when the children were taken to China to be raised by their Chinese relatives. The 1933 passenger list of the *Adrastus* included Samuel Lowe, merchant, forty-two (he was really forty-four); Swee Yin Ho, thirty-two, married woman; Barbara Hyacinth, one; and Anita Maria, five weeks. This was my grandfather's final return to China.

Aunt Adassa and Uncle Chow Woo were fascinated by these documents. The websites and archives that I had accessed were, like Facebook and Twitter, not available in China. Neither were records that could assist them in linking facts concerning their overseas Chinese relatives. I don't speak their language, but their facial and body movements, their surprise and joy at seeing these records, required no words. My aunt pulled out her birth certificate, an exact replica of what I handed them. In fact, Aunt Adassa and all of my mother's living siblings have their birth certificates. In the 1970s, when regulations loosened a bit in China, some family members applied to Jamaica's registrar general for their birth certificates and received them in China.

At some points we were all talking and our various English-speakers were having a hard time figuring out whom to translate. It hardly mattered. This was not the loud, boisterous excitement that sometimes occurs during Williams family reunions, when my brothers and I bring together our children and grandchildren. Here, with the dignity and reticence of my uncle and aunt, the tone was of contained, complete joy. We would look at each other and smile. That was enough. Then the conversation would begin again.

"Uncle Chow Woo," I asked, "do you have pictures of my grandfather? I have photos of my mother for you." My uncle retrieved a manila envelope containing two old, yellowed photos. I looked down at one photo and saw a solemn young Chinese man in a light Western three-piece suit and a dark tie with his young wife, wearing a light blouse and long skirt. They sit a decorous distance apart; the wife has a baby on her lap; and two little boys, dressed in short pants, stand solemnly by their parents. It was a portrait taken in China of my grandfather, his wife Swee Yin, and their boys Chow Ying, Chow Woo, and Chow Kong.

I looked carefully at Samuel's face and saw a man in his late thirties, his hair carefully combed to the side, his eyes intense, his whole bearing very formal and still. I saw his young family and was struck by the knowledge that, when this photograph was taken, his oldest daughter, my mother, was somewhere in Jamaica. She would have been around twelve or thirteen, working hard, abused by her grandmother, wondering what had happened to the kind Chinese man who was her father.

I then turned to the second photograph, a formal close-up showing my grandfather as a gray-haired, solemn, handsome man, wearing not a Western tie but a Chinese collar. As I looked at my grandfather's face, a face I had imagined for much of my sixty years, I saw glimmers of a face that I had known all my life. I gave my uncle Chow Woo some photos of his half sister Nell Vera Lowe, and he looked at her with tenderness in his expression. Any misgiving that he might still have had about our family connection disappeared as he smiled and passed the photo of my mother, age twenty-seven, to his sister Adassa. He said something in Hakka. My cousin nodded and smiled. Then he turned to me: "He says she looks like him. She looks like their father."

Yes. I know she does, Uncle Chow Woo. My mother, Nell Vera Lowe Williams, is finally claimed. No documents are necessary. Her umbilical cord was reattached in August 2012, in the Shenzhen Marriott hotel.

ADASSA TAKES CHARGE

THE PHOTOGRAPH IS VERY SIMPLE: TWO HANDS. ONE HAND IS OLD AND gnarled, the color of faded parchment, with a longish nail on the little finger. The other is younger—not a child's, but the hand of someone in middle life, from thirty to sixty—and a deep golden brown. The older hand has settled lovingly on top of the younger one, embracing it with intensity, as if to ensure that although years were lost in the past, years in the future will never be lost again. One hand is that of my aunt Adassa, my mother's half sister, who was born in the same year as my mother, 1918. The other hand is mine. It was August 2012 and we had just met for the first time. In a week I would celebrate my sixtieth birthday. Aunt Adassa was ninety-three.

There were seven children, Uncle Chow Woo had said via Yiu Hung in our earlier e-mails: Samuel Lowe had seven children, but none of them was named Nell Vera Lowe. I replied that when my mother was about sixteen she went to her father's shop, where two uncles told her that Samuel Lowe had returned to China and was not coming back. I also mentioned that I was planning a trip to China. I raced to my e-mail in the morning and saw that there was a response from China.

Dear Paula

    My name is XUHUI (Henry) LUO, I'm grandson of Chow Woo, who is one of the Samuel Lowe's sons. Chow Woo is living in Shenzhen at the moment. The Samuel Lowe is my great grandfather. I just talked to my grandfather and he explained to me the story of Samuel Lowe. He would like to meet you in Shenzhen on 14 Aug.

And now, here we all were. My Uncle Chow Woo, my Aunt Adassa, my mother's two half siblings sitting quietly and formally in the lobby of this comfortable Western-style hotel. I looked at their faces and in each one I could see something of my mother's beautiful face. I saw my mother's smile on the face of eighty-four-year-old Uncle Chow Woo, and her eyes in Aunt Adassa's eyes. Children and grandchildren were there. My two younger cousins, Gary and Stanley, who were in their twenties or thirties, whip-smart, and good at languages, served as translators.

    I discovered that our family name has many variations—Luo, Lowe, Law, Lo—depending on which language the Chinese character was translated into. Written Chinese is a language of characters, none of which resembles the Western alphabet. These characters can be converted to Western letters—romanized—in various ways (one system of romanization is called pinyin). In 1905, when my grandfather boarded the ship in the Hong Kong harbor, he would have told the British ship's clerk that his name was 罗定朝 and would have pronounced it "Lowe Ding Chow." The clerk heard a name recognizable to his ear and spelled it "Lowe." (And the clerk decided to name this laborer "Samuel," so my grandfather's name was recorded as "Samuel Lowe.") If someone who spoke another language heard my grandfather's name, that person could have spelled it "Luo." Yet another could have spelled it "Law" or "Lo." These variations in spelling using Western letters have little meaning to the Chinese, because in China the name Lowe is written with the symbol 罗.

I learned that of Samuel Lowe's children in China—including my mother, there were eight of them—Adassa, my mother, and Adassa's brother Gilbert had Jamaican mothers. Swee Yin wanted one of her own children to have the honored place of firstborn son, so she had told Samuel to leave Gilbert, the true firstborn son, with his mother, Emma Allison, in Jamaica. Adassa never saw Gilbert again, but she longed for him all of her life—a longing that I can well understand. And she never knew my mother existed.

This led to a bit of confusion.

Aunt Adassa could not understand that there was a sister who preceded her, that Nell Vera Lowe was her father's eldest daughter and that I am Samuel Lowe's granddaughter from that child. She could not imagine that her father had another liaison, let alone one that took place at the same time as the liaison with her mother. This was simply unfathomable to her.

Aunt Adassa was convinced, instead, that I was her brother Gilbert's daughter. Gilbert she knew about. She had grown up with the knowledge that she had a little brother who was left behind in Jamaica and that he, like her, was half Black. She'd concluded that this Black woman from the United States was Gilbert's daughter. And so it took some time and repetition for her to understand who I was and how I came to be her niece.

Adassa held my hand and said, softly but forcefully, in Hakka, "Now after so many years that our blood was apart, we can never lose each other again."

"Yes, Aunt Adassa," I said through my tears. "Never again."

Then, with unexpected vehemence she said, "You must bring everyone here."

Everyone?

As emotionally overwhelmed as I was, the highly developed part of my prefrontal cortex responsible for logistics immediately sprang into action. I knew that by *everyone* she would mean my brother Elrick and his son Chan; my brother Howard, his three daughters, and his grandchildren; my husband, Roosevelt; our daughter, Imani; and Imani's son Idris. I imagined bringing at least seventeen people from all over the United States, and organizing their complicated schedules. Of course I could do it. It was August 2012. Perhaps we could come in June 2013. A nice early-summer family reunion and vacation in China.

"Of course," I replied. "Of course I will bring everyone."

"When?" she asked.

How absurd I was to be thinking in terms of many months when I was talking with a woman who at the age of nearly ninety-four had already outlived most of her peers. It could not wait. We would make it a Christmas trip.

"OK," I said. "Before the end of the year."

"When?" she asked again. She was an elegant negotiator, my aunt Adassa—determined, focused, and not letting go until she got where she wanted to be. I admired that and I could feel the kinship between us strengthening. I have been known as a pretty good negotiator myself.

"OK," I said. "December."

"OK," she said.

She squeezed my hand and looked at me, her face infused with tenderness and happiness. She was reserved, and she was embracing. She was someone whose very way of being was familiar to me. I could feel the presence of my mother surrounding us all.

Since December 23 was going to be my aunt Adassa's ninety-fourth birthday, I could see that I had finally given her the correct answer.

How my mother would have loved to know Adassa, so serene and clear. They seemed to possess the same fierce maternal energy and produced similarly accomplished children. Adassa enjoyed a happy marriage, but when we finally met, she had been without her husband for thirty-one years. In 1979 her sixty-three-year-old husband had died, leaving her with five children.

The two sisters—Nell and Adassa—never knew each other, never knew of each other's existence. They were separated by continents and oceans, and each had experiences that the other could hardly imagine. How could Adassa have envisioned my mother's life in Harlem—the poverty and the danger that surrounded us, the occasionally violent encounters with my father, or even the ambience of a city like New York? And how could my mother imagine being taken from Jamaica to China as a very young girl, having three Chinese brothers and two Chinese sisters from her father's marriage to Swee Yin Ho, but being separated forever from her infant brother.

Adassa was a child when she left Jamaica, and yet she still had her Jamaican passport and her birth certificate. Her father and his Chinese wife took her and her three brothers to China in 1927 and left them there with his wife's family. They returned to Jamaica and had two more children: Aunt Barbara Hyacinth and Aunt Anita Maria. Aunt Adassa has only the faintest memories of the island where their father lived and worked. She was raised to be a productive, dignified Chinese woman; but she and Nell were blood, and as an old Hakka saying has it, bloodline is the root and culture is the soil.

Aunt Adassa held my hand. She was as much clinging to the present moment as she was ensuring the future when we would all be together.

GILBERT'S JOURNEY

AUNT ADASSA'S YOUNGER BROTHER GILBERT REMAINED IN JAMAICA. SAMUEL Lowe's wife refused to bring him to China, refused to accept him as the firstborn son, a status that she insisted must be conferred on her own oldest son, Chow Ying. This was so even though she had accepted Adassa as her daughter. The reasons were unarticulated but seem obvious: in Chinese history and social policy, daughters were not as important as sons, especially firstborn sons. In fact, the importance of the first son in Chinese culture is overwhelming. Gilbert looked more Jamaican than Chinese, so there would be no way to give an impression that Swee Yin was his mother. This would mean that some other woman, who was not even Chinese, would have the position of honor as the firstborn son's mother. There was no way that Swee Yin would relinquish her own status and her son's status for Gilbert.

So Gilbert, like my mother, remained in Jamaica. Unlike my mother, he maintained contact with his father's family: he was left in Saint Ann's Bay to be raised by my grandfather's nephew Leslie Lowe, alongside Leslie's nine children. Gilbert never visited China and never knew that he had a half sister living on the same island. Samuel Lowe fathered eight children, and Gilbert fathered ten—of whom most remained in Jamaica but some moved to New York. Gilbert died fairly young, and I had no idea of his existence until I began searching for my family and encountered someone who knew someone who knew someone in Jamaica who knew Gilbert's family.

I had no idea, until I was sixty years old, that there was a Loraine Lowe, a first cousin who lived in Brooklyn with two of her sisters, Annie Marie and Andrea. For most of my life, Loraine and I, our grandfather's descendants, had been separated not by the ten thousand miles between America and China, but by no more than fifteen miles on the subway. I first learned of Loraine's existence about a week before I went to China and had no time to talk with her then. But after meeting Aunt Adassa, after seeing how intent she was on finding some connection to her younger brother, after holding Aunt Adassa's hand as she insisted we return as a family, I knew that I had to talk to Loraine and get her to come to China.

I called Loraine from Shenzhen, and she answered the phone. I introduced myself. She seemed hesitant but I was fired up by having just met my family, and I felt a sense of urgency because Uncle Chow Woo and Aunt Adassa were so old. I had to press on. I told her that Aunt Adassa and Loraine's father, Uncle Gilbert, were full brother and sister. Our aunt, I told her, was almost ninety-four and wanted to see her brother's children, to meet them, to hug them. I told Loraine that I had promised Aunt Adassa that everyone would come to China in December, and I asked her: How many people are in your family?

Loraine seemed bewildered. China? In five months? She didn't believe that her family could make such a trip on such short notice.

"Loraine," I said sternly. "Aunt Adassa's a really old woman. She thought I was you or one of your sisters. It's *you* who she really wants to meet—not me. She never even heard of my mother until a week ago. It's *your* family she has been searching for, *your* family she has dreamed of meeting. Make it happen. Talk with your brothers and sisters and, even if not all of you can come, some of you *must make this trip.*"

I asked Loraine to start working on it and said I would call her when I returned to the United States in a few days. Later, over a drink, Loraine told me that she would never have believed such a call, such a plan, such a trip could drop out of the sky like that.

But she didn't know the power of our grandfather and my mother, pushing me from the other side.

MY HEART WAS
SOMEWHAT DISTURBED AND
I FOUND IT VERY STRANGE

MY UNCLE CHOW WOO IS NOW THE HEAD OF THE CHINESE LOWE FAMILY. HE
is the most esteemed elder, the third son of Samuel Lowe and
Swee Yin Ho, and my precious uncle who gave me my Chinese name. He was born in November 1925 in Jamaica. "Everyone
knew that my father had two other children," he said. "A sister named
Adassa and a brother named Gilbert." The Lowes knew that these
children's mother was not Chinese but Jamaican.

Two other children . . .

My mother never existed in their minds or in their family story.
(I suppose that there is a parallel in our experiences, because these
seven brothers and sisters never existed in my mother's narrative, her
sense of self, or her sense of place.) Then, all of a sudden, everything
changed for my sedate, elderly uncle, whose life had had its ups and
downs, its share of trauma and unexpected events, but whose old age
had been calm and prosperous—all of a sudden, doors were flung
open to a world he had no idea existed. He learned that there was a
niece, and that she was searching for relatives.

And that this niece was a Black American.

And that this niece was the daughter of his oldest sister, who he never knew existed.

And that in his already large family there were brothers and children and grandchildren and great-grandchildren, who were various shades of brown and had various shapes of eyes.

I had been so absorbed in my own experience of this discovery that only now, when I am somewhat removed by time and distance from all the excitement, can I contemplate the cataclysm that our appearance must have been in the lives of my Chinese family. I received the e-mail from my uncle Chow Woo's grandson telling me that Chow Woo would like to meet me. But I never realized that behind the warm invitation there was some skepticism, or at least some concern.

Before my uncle and aunt and cousins went to Shenzhen to meet us, my uncle spoke to his younger brother Chow Kong in Guangzhou. He described having doubts, being confused, "because we had no knowledge of this older sister. The first thing that came to our minds was: This is very odd. How can it be that another older sister existed? Because this is our older sister, her name is Adassa. She was taken home to China from Jamaica. But another older sister? My heart was somewhat disturbed and I found it very strange."

Of course he must have been unsettled. What a tremendous shock it must have been for him to discover, in effect, that there were *two* other women in his father's life: my mother—his new half sister—and the woman who bore her. That Samuel Lowe might have had other liaisons was not surprising. He was thirty-two when he married and he surely had not been celibate before Emma Allison entered his life. A previous mistress was one thing, but how could a man like Samuel Lowe, a Hakka, for whom family was everything, possibly have had another child and have kept that child a secret from his sprawling Chinese family? Chow Woo even called relatives in Canada to find out if they had ever heard about this mysterious sister.

My uncle urged me to bring the Jamaican birth certificate and any other documents that I might have. He and Adassa still had their Jamaican birth certificates, and in 1990 they had both applied for Jamaican passports, hoping to make a pilgrimage to their father's country, the place of their birth, and perhaps even visit his former store and home. When we met, we compared the birth certificates and the other documents. I looked at my grandfather's English and Chinese names: Samuel Lowe and Lowe Ding Chow.

Now that we were all together, it seemed they accepted me without any hesitation. My uncle Chow Woo embraced me. Aunt Adassa knew that we were blood relatives. What reinforced my uncle's sense that we were family was less evident; it was a matter not of documents but of shared sensibilities. What reinforced it still further was my quest. My uncle admired my perseverance, my resolute effort, and saw this as an example of the Hakka tradition of the primacy and importance of family.

LOWE SHUI HAP

OUR FAMILY VILLAGE IS CALLED LOWE SHUI HAP, AND AFTER MEETING AUNT Adassa and Uncle Chow Woo in China, I knew that I had to go to this village and see where our family came from, I wanted to walk along the streets where my grandfather had walked. Lowe Shui Hap means "New Crane Lake," and the dwellings with that name are now a state-run museum. Once it was the name of the walled building that was the family compound.

At our first meeting in the lobby of the Shenzhen Marriott, I asked my uncle Chow Woo if he would be able to join us when we visited Lowe Shui Hap the following day. He was eighty-four years old and seemed a bit tired after the hour we had spent together, after all the photos and the revelation that he had a sister and the long journey that had brought him to this reunion. As exhilarated as I was, I could also feel emotional exhaustion threatening my equilibrium. What must it have been like for him? I imagined that beneath his serene face and his mild manner, consuming, unaccustomed emotions must have been swirling. He, the middle son of Lowe Ding Chow, the oldest surviving son, and now the family patriarch, had learned much about his father on this steamy August afternoon, while sipping tea with this tall brown-skinned woman from America.

How could there be another oldest sister? Adassa, or "Big Auntie," had always been his big sister, the surrogate mother who had cared for him, and scolded him, and reminded him of his duty to family—just as their father had instilled it in them, and just as she had always observed it, placing her half brothers and half sisters first. Sitting next to him on the couch in this hotel lounge, Adassa was serene. Her face showed no surprise, no disbelief. She is his eldest sister. Now, there is another eldest eldest sister? Nell Vera Lowe: how had their father not told them of this firstborn girl child of yet another mother? No one was concerned about knowing that Lowe Ding Chow had two Jamaican or "local" partners, as the Chinese referred to women like Albertha and Emma, unmarried legally but wives nevertheless. In China—and in Jamaica—the Chinese shopkeepers had as many "wives" as they could afford. And Chow Woo's father, Samuel Lowe, had been prosperous. In his shop in Mocho, Emma Allison was as much the shopkeeper as he was. In his Kingston shop, Albertha had been his partner and lived there with him and their daughter. And both women bore his children: Nell, Adassa, and Gilbert. Chow Woo now pondered. But why not tell his children born to both Emma and Swee Yin that they had another sister, his firstborn daughter?

Uncle Chow Woo thought about it and concluded that, since their father had been unsuccessful in his search to find little Nell, he had kept this part of his life from his children. As my mother's uncles told her, from 1921 until 1933, Samuel Lowe never stopped looking for Nell, but why talk about such sadness? He just didn't tell his children there was another partner and another daughter and another, truer "Big Auntie."

I think my uncle's look of fatigue came from his effort to decipher this untold story about his unknown sister—a sadness born of lost love, lost family, lost dreams; another unconnected and unembraced

Lowe. Nell, like their brother Gilbert, had been lost to Chow Woo, to his brothers, and to his sisters. "She looks like him," he said, almost mesmerized by her photo. She looks like him.

"Uncle?" I asked. "Uncle?" as I gently touched his forearm, hoping to lead him back to me from the labyrinth of his thoughts. He turned to me and smiled. Not the eyes-twinkling gaze I've subsequently come to expect whenever my dear uncle looks as me, but a resigned and accepting look that said, "I'm here. I know who you are. You are a Lowe and I'm your uncle. You are my family. Don't worry." I read it on his fatigued face. He'd thought it through and he was now unwavering.

He had impressive discipline and was clearly happy. His smile was one of rich contentment. When I asked if he wasn't too tired for a visit to the village, he straightened and looked astonished at the thought that he would not join us, indeed not be our guide on this pilgrimage.

"Of course!" he said, his face animated with passion. "It's *my* village. I grew up there," he added, patting his chest to emphasize each word. "And I am taking you there to see where your grandfather was from. The people at the village, they don't know *these* people," he said with a wave of his hand to include the well-meaning younger generation sitting at the table. He looked me in the eye, and his face was animated and youthful. "They know *me*. It's *my* village," he said. "And *I* will take you!"

Aunt Adassa smiled with an older sister's indulgence for an energetic younger brother. But I could see the fatigue on her face and I felt a pang of anxiety as I thought of her ninety-minute drive the next day to her home in Guangzhou. We hugged and kissed and said our farewells. As Aunt Adassa and I embraced, I could feel how old and fragile she was. I realized that I might never see her again. She was just one month younger than my mother, who had been dead for six years. As she walked away, her son on one side and her daughter on the other, I noticed her slow, deliberate pace, her slightly hesitant gait as she clung to her walking stick. I wanted to cry out after her, as much a blessing as an entreaty, "Aunt Adassa, we'll be back in a few months in December. You have to be here. Stay alive, Aunt Adassa."

In contrast, my uncle walked away with a spring in his step: strong, assured. There was nothing wobbly or feeble about him. He repeated the plan for the next day, just in case I had missed anything: We would meet at a rest stop along a highway and then he would take me to Lowe Shui Hap. It was early evening and I was to meet Marcia later for dinner. I went to my room and stared at my grandfather's photos.

The following morning, cousins picked us up at the hotel and we set off on our field trip. After the thirty-minute drive from Shenzhen to Lowe Shui Hap, we arrived at an imposing, enormous two-story wall, with a few small arches carved out for doors. Two cannons supported on several blocks pointed outward (cannons were invented in China). The vast scene looked medieval. The village, like Shenzhen, had once been a small borough. Deng Xiaoping, the forward-thinking chairman of the Communist Party's Central Committee, decided to establish Shenzhen as a mainland center of business and commerce, an alternative to the British colony Hong Kong, just a few miles away. Shenzhen grew up around Lowe Shui Hap, enveloping it, but the government decided that it should remain an intact ancestral Hakka village and converted it to create the largest Hakka cultural museum in China.

Lowe Shui Hap was a revelation. We walked into the family compound, "New Crane Lake Dwelling," through an ornate traditional gate with a high threshold and through an inner doorway. A plaque over the lintel read "Da Fu Di"—"The Home of a High Government Official." The original wooden plaque had been authorized by the Qing emperor of the day. Another plaque, over the entrance to the village, was the Lowe family motto: "Prosperity, Family, and Education." I had my cousin repeat those words and wished—not for the first time during this trip, but perhaps most intensely—that my brothers and mother were there to see this stunning affirmation of the three priorities of our small Lowe-Williams family in Harlem. Could there be a more concise summation of what our family lives by?

The dwelling was completed in 1817, about five generations after it had been started by a sixteenth-century Lowe, Luo Ruifeng. Actually, the Lowe family clan goes back 153 generations if, like the Chinese, we calculate about twenty years for each generation. The documented history of the Lowe family takes us back more than three thousand years, to about 1000 BC. The museum has a computerized "virtual book"—you wave your hand over the screen, and pages turn—that recorded the entire Lowe family, generation after generation.

I waved my hand over the legacy book called a *jia pu* and kept asking my cousins, "Where is my grandfather in this book? Where is he?" I had mobilized the family to try to find his name, and my cousins were scanning the pages, but so far in vain. Uncle Chow Woo stepped forward in his quiet, commanding way and told my cousins to back off. He waved his hand and pointed at an entry. "Baba," he said—the Chinese word for father.

My uncle then led me to a wall where a meticulously prepared family tree was displayed. He pointed to a spot on the tree and became more animated. "This is my father," he said. "And this is our line." There were many names surrounding that line. "Are women represented?" I asked, somehow already knowing the answer.

He looked surprised, as if I had asked whether the family pets were on the tree. "No," he said, shaking his head. "No women." This was a record of men through the generations, the men who carried on the name Lowe. Painful emotions collided within me: love of my uncle; pride in our line; outrage that women did not signify; penetrating sorrow that my mother, once again, was left behind; determination that something must be done for my mother, for my children and grandchildren and great-grandchildren for the next 153 generations.

I wanted some public recognition, in this village, that my grand-father's dream—education, prosperity, family—was fulfilled with my mother, with my mother's children, with her grandchildren and her great-grandchildren. There must be a place where generations two, three, four, and onward continue after Nell Lowe. My situation was poignantly ironic: here we are, descendants of slaves; the best we might be able to do would be to go back two, three, or possibly four generations; and I refuse to be—I fight, I resist being—narrowly defined as a preslavery and postslavery Black person. When I saw this family tree, then, I thought, "I knew it. I knew it."

I looked at my uncle in astonishment and imagined the power of all those generations coming to bear on the two of us. Many African Americans don't know where we came from before we were captured, forced onto auction blocks, and sold as chattel. To travel to Lowe Shui Hap to see my family's ancestry and realize that it predates Christi-anity by more than a thousand years was almost surreal. Indeed, it was life changing. Even so, Africans predated the Chinese and their documented history. The African man was the first man.

I began to comprehend a new way of looking at my place in the universe; standing there and seeing the thousands of names of my ancestors, I realized—almost with the force of a religious conver-sion—that my being is not a stray occurrence in the universe; it fits into a much larger scheme. I cannot express this as "predestination," because that word is freighted with old religious biases, but I do feel that I didn't just happen randomly. Like my ancestors from one hun-dred, five hundred, two thousand years ago, one day my grandson will grow up knowing the names and legacies of his ancestors, as will his children, and their children. Uncle Chow Woo is the 150th generation; I am the 151st generation; and my grandson, Idris, is the 153rd generation.

At the start of the visit, Uncle Chow Woo took me to a temple, where he lit a fistful of incense sticks. He handed three sticks to me, keeping about ten for himself, and walked me to the altar. Motioning, he told me to climb one ladder while he climbed another. We ascended about three rungs, bowed three times, and said silent prayers to our ancestors. I prayed to my mother and my grandfather. I said a fervent prayer of thanksgiving to the unknown powers and people that had led me to this moment.

We then climbed down and put the incense sticks in a vessel, where they would continue burning as we walked through the village. My uncle walked with the sure-footedness of a man familiar with and in charge of this place. In fact, I saw someone make the mistake of trying to offer him an arm for stability as he crossed a high threshold. He waved the person off—tactfully but firmly, as if to say, "I am fine. I don't need help. Thank you."

Only security guards live in Lowe Shui Hap today, but it housed members of the Lowe family until 1996. My uncle showed me the gathering place for the elders, who are revered in Hakka culture. He took me to his father's home, where Samuel's two brothers' families had occupied the adjoining houses.

I met a somber man about my age whose name is Xinrong. He and I are second cousins, since his grandfather and Samuel Lowe were brothers. Uncle Chow Woo said Samuel Lowe was closest to Xinrong's father, Sih Chiu. Xinrong is a regional village leader and as such has oversight of the museum and is the secretary general of the Communist Party for the region. He spoke to me in Hakka in a way that conveyed his authority—gently, but it was there. He said that we were like brother and sister because we are of the same generation. We spoke about his grandfather and mine being brothers. He initially pointed out that his rank was above mine. I asked the year of his birth, and my cousin translated it: 1956. I smiled kindly at Xinrong, pointed to my own chest, and said simply, "1952." When the year of my birth was translated for him, he smiled broadly and assumed a more respectful, almost reverential attitude: "Big cousin. Older cousin." Such is the importance of birth order and rank in China. No wonder Gilbert was left behind.

We headed to my grandfather's house and I stood at the door. The house was small. During the late 1930s, Samuel had built an attic for the children to sleep in; from the front of the house, it extended from the interior to the outside. The Hakka writing bracketing the front door urged all who entered to "Go and make your fortune, become prosperous, and return home." I could hear my mother insisting that we all become wealthy. She had learned this lesson from her father as thoroughly as she had learned to count in Hakka.

The whole experience was overpowering for me, leaving me with much to process and integrate into my worldview. In a few weeks, my world had been expanded by another continent—a huge continent where one corner houses my family. I have a place, a village, elders, cousins. They know me and they know my name. They have my mother's pictures and share her blood. I went home to my grandfather's house. I was in his kitchen and in his bedroom. I listened for him in Uncle Chow Woo's voice and I imagined he had that same self-assurance, authority, and commanding presence in a room.

I am Hakka.

I am Lowe Ding Chow's granddaughter.

I am Nell Vera Lowe's only daughter.

I am a Lowe.

PART SIX

SAMUEL LOWE'S TWENTIETH CENTURY

Our greatest glory is not in never falling,
but in rising every time we fall.

—CONFUCIUS

A JOURNEY OF
TEN THOUSAND MILES

IN 1889, THE MAN WHO BECAME SAMUEL LOWE IN JAMAICA WAS BORN DING Chow (or Chiu) Lowe, in the village of Lowe Shui Hap—"New Crane Lake"—in Longgang, Shenzhen, Guangdong Province. Samuel was the middle of three sons of a brilliant entrepreneur, Yuxiu Luo. The oldest boy may have enjoyed the important "first son" status, but he was a libertine and a spendthrift who lacked his father's discipline and drive.

Samuel was different. A born entrepreneur, he shared his father's business sense and ambition, but in a culture as rigidly hierarchical as this, he could not escape his status as second son. My grandfather was born at the end of the Qing dynasty, the long, last Chinese dynasty that extended from 1644 until 1912. The society into which he was born had five classes, or estates. The top three were the ruling class, the tiny aristocracy, and the educated class who had been to universities and held degrees. The commoners or working classes were divided into the *liangmin*, the "good" commoners, who made up most of this group and included some scholars, farmers, businessmen, and artisans; and the *jiangmin*, those who were ignoble, or "mean," such as slaves, servants, actors, prostitutes, and some low-level government employees.

This was a period of mobility and turmoil in China. The population had grown to 300 million in the nineteenth century, and with that growth of population had come increasing movement. The government encouraged travelers, of whom some were transient and others permanently relocated to other regions of the country.

My grandfather's family were probably *liangmin*. Samuel attended school for only two years before there was pressure for him to begin work. He went to the villages of Lowe Shui Hap and Loong Kong, looking for employment. He wanted to work in a small shop, but there were none that would employ him, so it was his own responsibility to do something productive. Already, some of his relatives in Lowe Shui Hap had gone to Jamaica.

Before 1891—when the Qing reversed their policy—the Chinese government discouraged emigration, worrying that too many laborers would seek their fortune in the West. The first Chinese in Jamaica had emigrated in July 1854 in defiance of the prohibitive policy. They went initially to Panama as indentured laborers, working on the railroads. When yellow fever struck in 1854, 205 of these Chinese workers demanded that they be permitted to break their indenture and leave Panama to escape the epidemic. They were given permission and went to Jamaica in two ships, in 1854. Over two hundred Chinese arrived there, but for three-quarters of them it was too late; they died in Jamaica of the fever. Only the hardiest survived—fewer than fifty—and three of them became the fathers of Chinese retail in Jamaica, opening dry goods stores that were the precursors of the legendary Chiney shops.

Other Chinese immigrants arrived in the 1860s. They were also indentured workers who had three-year contracts with American companies in Trinidad and Tobago, working in the sugar, banana, and coconut fields. When they had fulfilled their contracts, they gravitated toward Jamaica, where they saw greater possibilities for creating their own businesses. At first slowly and then, with the arrival of eight hundred Chinese immigrants in 1888, much more quickly, the Chinese community grew.

I don't know if any of my relatives were among the eight hundred who arrived in 1888, but in any case Samuel Lowe did have some family from Lowe Shui Hap already in Jamaica when he set sail to seek his fortune. He was leaving a China that was still reeling from the effects of the Boxer Rebellion, which apparently arose in Shandong in 1898. This rebellion, which has also been called a "cryptonationalist movement," involved mostly poor peasants, who had been organized and given a certain standing by a combination of martial arts and invulnerability rituals. Most of the activity occurred in northern China, where the group initially harassed foreigners, especially foreign missionaries, and Chinese converts.

The Qing court was not at all unhappy with this group. Indeed the Qing considered the Boxers useful as a means of ridding China of foreign influence, but there was also some concern that eventually the antipathy against missionaries and other foreigners would be turned against the dynasty. As a preemptive step, the empress decided to work with them and even provided modern weapons from the imperial army to augment their supposed magical powers when she ordered them to massacre all foreigners in June 1900, and declare war on all foreign powers.

Many foreigners were brutally murdered, and the legation quarter in Beijing was under siege. Western and Japanese troops fought their way in from the coast, arriving in time to save foreign diplomats and other foreigners who had taken refuge there. Then the Chinese officials themselves became divided. Some worked with Boxers, while others protected foreigners in their jurisdictions. The situation was becoming uncontrollable, and finally, in 1901, an alliance of eight European nations moved in to quash the rebellion and restore order. The humiliation of having Chinese internal affairs managed by external imperial powers lingered for years.

Samuel would have been eleven years old at this time, the middle son in a family that worked to make ends meet. He may not have known much about what was causing the political tensions or why there was a drought. But he knew that he had to contribute; he had to help put food on the table and try to make his family's life more secure. When Samuel was around fifteen—we cannot be exactly sure of his age—he traveled, for over two months, the 9,583 miles from Guangzhou to Kingston, Jamaica, to become a laborer. The ship he took must have been crowded with other boys and young men eager to make a fortune in this faraway island. He might have stopped in Panama, since the Panama Canal was about to be built and laborers from Asia were gathering there to be part of the adventure.

But instead he became a part of the great transition from slavery to free labor that occurred during the late nineteenth and early twentieth centuries in the Americas and the Caribbean basin. The system of indentured servitude prevailed in that region, and we believe my grandfather began his life in Jamaica in 1905 as a teenage indentured worker with a three-year contract. If so, his first job would have been cutting sugarcane, exhausting work in an industry that was then already on the decline. The conditions weren't as punishing as they had been in mid-century, when some workers died of starvation, but they could not have been very comfortable or lucrative.

Nonetheless, he was so frugal that by the time his indenture was completed, he had saved enough money to be somewhat independent. A relative of his had a small convenience store—a "penny shop"—in the mountains, and Samuel went there to work with him. To call the place a shop may be an overstatement. It was actually only a stall set into a mountainside, but it sold rice, cornmeal, salt meats, and flour to the local Jamaican population, and for Samuel Lowe it was retail. Finally he was able to work in an occupation that had been out of his reach in China.

After a while, his relative had enough of rural life and decided to move. My grandfather, who had assiduously saved nearly every penny he earned for just this moment, bought the small store from him, took it over, and worked all by himself on the mountain. Eighteen years old, Chinese, all alone, gradually he befriended the people in the community, extending credit when they needed it, making special orders if they asked. Eventually, he hired another Chinese, who provided both assistance with the work and companionship in the lonely world of the Jamaican countryside.

And this is how it all began for my grandfather: penny by penny, with long hard days and nights at work, enduring loneliness that could have broken a less disciplined man. Gradually he was able to save enough money to move out of the mountains and establish a shop in Mocho, then another in Saint Ann's Bay. He sent for both of his brothers to join him in this prosperous business. The climate reminded him of their village in southern China, Lowe Shui Hap, and he was eager to have the three of them together build a legacy for their parents and other relatives. The eldest son, Hin Chiu, wasn't suited to the hard work and entrepreneurship that it took to run Samuel Lowe and Brothers; after a time, he returned permanently to China. But Hin Chiu had a son named Leslie who was born in Lowe Shui Hap but then was taken by his father to Jamaica. When his father went back to China, Leslie remained behind in Saint Ann's Bay and was raised by his uncle Samuel and, after Samuel married Swee Yin Ho, by his aunt as well.

In ten years, my grandfather had built successful businesses. He was ready to have a wife, a traditional Chinese bride with whom he could start a real family—in contrast to his improvised relationships with Emma Allison and my grandmother, Albertha Campbell. The marriage was arranged in China, and Swee Yin Ho was sent to Jamaica to meet her new husband. They were married shortly after she arrived, and the wedding was announced on page 2 of the *Daily Gleaner* of December 23, 1920:

WEDDING ANNOUNCEMENT
The wedding of Mr. Samuel Lowe of Mocho
and
Miss Ho Shui Yin,
daughter of Mr. Ho Chit Sang, of Kingston,
will take place
At the Kingston Parish Church,
on Monday 27th instant [*sic*] at 10:30 a.m.
By the Rev. Page.

I wonder if either Emma Allison or Albertha Campbell read the
announcement.

SECONDARY MIGRATION

MY UNCLE CHOW WOO WAS BORN IN JAMAICA. IN 1927 HE DEPARTED TO China with his parents and three siblings—Adassa, his older brother, and his younger brother. In addition, four other children traveled with them. They were the children of friends who wanted to send them to China as well. Taking them to China was an important process, designed to ensure the children's integration and assimilation into Chinese society and their fluency in the language. These children might be Jamaican-born, some even with Jamaican mothers, but their primary cultural identity was Chinese. For my family, it was Hakka Chinese.

The Lowe family saga might appear to be a unique story, but in fact it was typical of many Chinese Jamaicans at the time. Scholars refer to this as a "secondary migration phenomenon," in which many Chinese immigrants to Jamaica would, in turn, send their children back to China for years, to live with relatives or sometimes with strangers. The point was to encourage, shape, and reinforce a strong Chinese identity during the formative years. Parents wanted to prevent their children from completely assimilating into the Jamaican culture that surrounded them.

Chinese immigration to Jamaica took place in three phases. The first was in the mid-nineteenth century, when indentured laborers were brought to Jamaica from China. Many worked off their indenture, remained on the island, and then began to build small businesses. The second wave occurred during the late nineteenth century and the first part of the twentieth, and was motivated by the promise of the economic opportunity that Jamaica offered. The Chinese immigrants who arrived in that wave were more entrepreneurial. The final surge occurred at the end of the twentieth century, when a new group of Chinese entrepreneurs came to the Caribbean.

The "secondary migration phenomenon" flourished from about 1915 until 1937, when the Sino-Japanese War broke out and families decided to keep their children safe in Jamaica. Typically, the fathers were Chinese shopkeepers and the mothers were either their common-law Jamaican wives or their Chinese wives who had moved to Jamaica; my grandfather had children in both circumstances. Many children were sent to China with siblings, but once there, these siblings often did not stay together—they were raised separately by various relatives or acquaintances. One can imagine how confusing all this must have been for the children who were separated from their parents and one another and immersed in what was to them a foreign culture. For some children it was a trauma from which they never recovered. For others, like the children in the Lowe family, the time they spent in China was a brief intermezzo in a more stable personal history.

My grandfather, his wife, and four children—including their one-year-old son Chow Kong—traveled from Kingston to Guangzhou. When they arrived in Guangzhou, which is about seventy-five miles north-northwest of Hong Kong, they left Chow Kong with his maternal grandparents. Samuel Lowe had married well: Swee Yin's father owned a successful herbal pharmacy and had invested in real estate the money that Samuel and Swee Yin had sent back from Jamaica. As Chow Woo said, "This was the dream of all Chinese, to go to Jamaica and become wealthy, and be able to return to buy farms, land, and build houses back in China. That was their hopeful thinking."

Samuel and Swee Yin went to Lowe Shui Hap, where Samuel's mother still lived. Three-year-old Chow Woo and his older brother—five-year-old Chow Ying—stayed with their paternal grandmother there along with their older sister, Adassa. After celebrating Christmas with their family, my grandfather and his wife returned to Jamaica to continue building their capital and expanding their business.

They stopped in New York en route to visit Swee Yin's brother and his family.

While there, they heard some terrible news: fire had destroyed their shop in Saint Ann's Bay, Jamaica.

FIRE! FIRE!

I N MY QUEST TO LEARN AS MUCH AS I COULD ABOUT THE CARIBBEAN HAKKA, I began to read books, journals, anything I could find that offered a glimpse into the life my grandfather led and my mother missed.

One book I found is Paul B. Tjon Sie Fat's *Chinese New Migrants in Suriname: The Inevitability of Ethnic Performing*. His focus is on the former Dutch colony on the northeastern coast of South America, which is 1,739 miles from Jamaica. But on page 183, he offers a fascinating analysis of the experiences of the Chinese—and they were overwhelmingly Hakka—in Jamaica. I found his observations helpful. The specific timing is a bit off for my family; still, as I read this passage the atmosphere in which Samuel Lowe lived became more layered for me, and more ominous. I could feel the menace in a way that I had never before experienced. He is writing about shopkeepers, about my grandfather and my great-uncles.

The Hakka shopkeepers ran their Chiney shops, the general stores for their local communities. But the providers of many of the goods to stock the shelves were members of the white elite, so the Chinese were often the middlemen, between the white elite who controlled the supply and the African-Jamaicans who constituted the demand. "In the Caribbean region the most remarkable anti-Chinese sentiments were found in Jamaica," Tjon Sie Fat writes, "the scale of which was unique in the British Caribbean." He points out that in the twentieth century, violence against the Chinese in Jamaica peaked three times: in 1919, 1938, and 1965. The conventional analysis is that this violence was caused by inevitable tension and resentment between the shopkeepers and their African-Jamaican clientele—that is, by stratification.

But Tjon Sie Fat suggests an "alternative explanation that points to White resentment of Chinese social mobility." Because the white elite controlled the media at the time, they could fan the embers of African-Jamaican resentment by promoting "the image of Chinese as parasitizing the Afro-Jamaican population," and they "pushed for tighter restrictions on Chinese immigration and stricter regulations to prevent Chinese dominance of the retail sector."

Debates about the Chinese were a regular feature of Jamaican public conversations. In their book *The Story of the Jamaican People*, Philip Sherlock and Hazel Bennett describe some of the immigration restrictions that were imposed on the Chinese. In 1905 the government began to restrict the entry of Chinese into Jamaica. Immigrants had to apply for registration with the immigration authorities and they had to be recommended by some member of the local Chinese community who would provide what was in effect an affidavit assuring the authorities that the new immigrants would not become a burden on society. This affidavit also had to guarantee the immigrants' good conduct. An official permit was then issued, which the immigrant had to produce on arrival.

During the latter part of the nineteenth century and the early twentieth century a halfway house was established in Panama, where prospective immigrants to Jamaica could obtain clearance. They might have to wait for as long as five years or more. Sometimes the permit was not forthcoming, as unscrupulous agents, having taken money from sponsors in China as well as in Jamaica, had no intention of troubling themselves any further. The applicants stranded in Panama could not return home or they would lose face. Some took the only honorable way out and committed suicide. In 1910 the immigration law was further amended. It required Chinese to deposit thirty pounds sterling on arrival, to be refunded after one month. They were also expected to demonstrate written and spoken familiarity with at least fifty words in English, French, or Spanish and undergo a physical examination.

According to the 1911 census, the 2,111 Chinese residents of Jamaica were only 0.3 percent of the total population. But neither this fact nor the restrictions on immigration prevented the newspapers of 1912 from complaining about the "Chinese invasion" and demanding action. The *Daily Gleaner* quoted a Jamaican citizen as saying, "Can the authorities do nothing to let Jamaicans feel that Jamaica is still their home and strangers will not be allowed to elbow them out of what is theirs by right?"

A later incident gives some idea of how toxic the discussions could be. In 1935, the legislative council debated a law to restrict immigration by the Chinese. In one of the debates a certain Captain Cipriani said, "We cannot allow people from outside to invade this country for the purpose of seeking employment or getting employment when our own people here have no employment. Only two weeks ago the Lady boat brought in 25 Chinese. I believe she is in again this morning with 18. . . . We have to put our shutters up against those who put their shutters up against us."

Whether or not they were aware of the more nuanced social dynamics involved, Samuel and Swee Yin often referred to the necessity of having "two strings in their bows" during the peak of their wholesale business around 1927. They had always worried that the racial conflicts in Jamaica would escalate, so while they allocated some of their profits to expanding their existing Jamaican business, they put the rest of the profits into the safety net of properties in Guangzhou, in case things got so bad that they had to leave. Samuel's brother-in-law Jiquan began buying properties for them in China, and by 1930 they owned a tenement at 76 Longzang Street, which they rented to a printing company. They also owned a 100-square-meter bungalow at the foot of Guanyin Hill (Guan Yin Shan).

Samuel had expected that these properties would be needed only in the distant future. He had big plans for his shop in Saint Ann's Bay, Jamaica, with more goods to sell and more clients to serve. Christmas of 1928 promised to be his best year ever. Before he left to take his four children to live with their grandparents in China, he fully stocked his shop. Samuel had prepared and bought special supplies for the holiday season. He had relied on his brothers to manage the shop and the warehouse behind it. As my grandfather and his wife sailed from Hong Kong slowly back to Jamaica, they were unaware of the catastrophe that took place in Saint Ann's Bay.

The citizens of Kingston awoke on Christmas eve 1928 to see a headline in the *Daily Gleaner*:

"FIRE! FIRE" IS THE CRY AT ST. ANN'S BAY!

This news competed with updates on the stable though serious medical condition of the British monarch, George V (the king did not die until eight years later, in 1936). But commanding the top right-hand side of the front page was the story of the fire that destroyed my grandfather's shop.

BLAZE LEVELS SIX BUILDINGS TO THE GROUND

———

Water Is Scarce and Fire Brigade Arrives Full Fifteen Minutes
After Alarm

———

WOOD FEEDS THE FLAMES
Police on Hunt After Thugs
Out Bamboo Way Are Thus
Absent from Scene

———

STARTED IN RUM SHOP

(*From our correspondent*)

Saint Ann's Bay, Dec. 22—At 12:30 this morning the tranquility of this township was disturbed by the dread sound of "Fire! Fire! Fire!" coupled with the ringing of the market and church bells. In a very short while the street was a seething mass of men, woman, and children who hurried out of their beds to the scene of the conflagration.

The "who, what, when, and where" of the event were thoroughly covered. The correspondent described how a girl named Berryl Holiday, who lived nearby, saw the flames, discovered the fire, and sounded the alarm. The locals referred to the neighborhood where it all took place as Guenep Tree, after trees that are indigenous to Jamaica and produce a strange, tasty green fruit, somewhere between a lychee and a lime. David Walters, a businessman whose shop was near the area, raced to the scene and tried to help put out the fire, but it was already blazing out of control "fed by wooden structures of considerable age." The delay in the response of the fire brigade was offset a bit by the appearance of a Mr. Wood, who was the manager of Barclay's Bank and was in possession of a fire extinguisher.

One business and home after another was wiped out: the house of Mrs. Hay, the registrar of births and deaths; the shops and residences of Mr. Kings Lee; the tavern across from my grandfather's store; the home and shop of Mr. Morris; and the rum shop of Mr. W. H. Scott. But ground zero was "to the rear of the premises of Samuel Lowe and Bros." The men tried to put out the fire there, but "the shop premises of the Lowe Brothers was then a mass of flames and these men had to be kept immune from an attack on their clothing by the fire, the heat of which was almost unbearable, by a constant shower of water being poured on them while they applied the chemical extinguisher."

Finally the fire brigade arrived, the inferno was over, and the lingering question "Why?" hovered in the smoke. How could this have happened? The earliest reports suggested that the fire had been deliberately set. By December 26, the *Daily Gleaner* reported, in bold print, "Crime Suspected." Near the end of the story that enumerated the devastating losses came the observation, "It is abundantly manifest that incendiarism is at the bottom of this regrettable episode in the history of this township, and it is sincerely hoped that before long the police might be able to bring to justice the person or persons responsible for this very regrettable disaster."

In my grandfather's life, the disaster was far more than merely "regrettable." The total loss in Saint Ann's Bay was estimated at 18,150 pounds sterling, which in today's currency would be 926,000 British pounds or over $1.5 million.

In today's dollars, my grandfather's shop, dwellings, and stock were worth over $511,000. While the paper reported that the stock had been insured for 2,000 pounds sterling, it hadn't been. Add to the losses the fact that any goods not destroyed by the fire were then looted by the locals.

Who would have done such a thing?

The question has remained unanswered for generations. The superficial answer is that local toughs set the fire, and some were subsequently prosecuted.

When Samuel Lowe and his wife returned to Jamaica in January 1929, they returned to loss and destruction. They rebuilt, using funds the insurance company paid for the destruction of the building. But Samuel got nothing for the uninsured inventory, an inventory he had fully stocked before setting off for China with his wife and children.

I am not sure how my grandfather did it, but on Monday, January 21, 1929, about a month after the fire, Samuel Lowe's new business announced its presence in the world, in an advertisement, of course, in the *Daily Gleaner*: "We beg to inform our friends and customers that we have temporarily opened up our business at 24 Main Street St.

Ann's Bay, where our usual courtesies will be extended. We wish to thank you for your past support and solicit the continuance of same. Samuel Lowe & Bros. St. Ann's Bay."

Financially, the situation could hardly have been worse. Samuel had little insurance money, and it cost a great deal to rebuild his shop. The goods with which he stocked the shelves were on consignment, so Samuel owed the creditors. Add to that the fact that his customers mostly bought on credit. His business model was therefore not very secure. Even though he created the new small shop within a month, it had only a small fraction of the earlier goods, so his longtime customers simply went elsewhere.

My grandfather was determined to make things work. The small shop was only an interim solution. His plan was to create a much more modern, more sophisticated emporium that would have electricity. By the end of July of that year, he succeeded. The area that had been devastated by the fire had been largely rebuilt and now had, according to the paper, "quite a distinguished air about it. Nearly all the buildings have been rebuilt and before many weeks are passed will be ready for business. . . . Messrs. Samuel Lowe and Bros. have erected an elaborate two-story building to house their wholesale and retail business. They are now occupying the upper portion of the building. Nearly all of the new buildings will be lighted with electricity which will certainly add to the fine appearance they now possess."

But Samuel's vision may have exceeded the stark realities of the market.

And he could not have predicted that when his gleaming new store opened in July 1929, an economic Depression was about to overwhelm the world.

THE RETURNING WAVE

NO COUNTRY IN THE WORLD WAS SPARED THE GREAT DEPRESSION OF THE early 1930s, and the shock was felt in the small towns of Jamaica as well as the large cities of America, Asia, and Europe. In Jamaica export demand and production contracted suddenly, with export prices plummeting by 44 percent between 1929 and 1932. The deterioration and weakening of local political, economic, and social conditions, and of living standards, prompted many overseas Chinese to leave their countries of residence. This marked the start of what is called the "returning wave" of overseas Chinese at the beginning of the 1930s.

After the fire, my grandfather rebuilt his store in Saint Ann's Bay into a beautiful modern emporium with an exciting new feature: electricity. But the goods were on consignment and Samuel Lowe owed money to creditors. Because his customers were used to charging their purchases, he had to struggle to bridge the gap between money coming in and money going out. He was a genial man, who appreciated being surrounded by people and providing for them. On weekends, many customers would come to pick up their orders, and Samuel would invite them to sit down for dinner. Customers and suppliers would lounge in the front of the store, where other

Chinese shop owners would join them. They would catch up on the week's gossip, the local news, and whatever new products might be in the offing. Although this may have been a wonderful way to make friends, it was not the most prudent approach when business was precarious. And despite all these friendly faces, many of his previous customers had abandoned him after the fire and had gone elsewhere to shop. Furthermore, the Great Depression was causing a severe tightening of credit, in addition to which my grandfather simply had some bad luck. In January 1931, for example, his van became disabled when the tires blew out during a delivery of bread. While the driver went on foot to get help, according to the *Gleaner*, two boys broke into the van; they were last spotted eating some of the bread and carrying more away.

Samuel simply could not make the enterprise work. Those final months must have been agonizing for him and Swee Yin. What could they do? How could they make it work? Was there an option still untried? Finally, in November 1931, he could carry on no longer.

A notice of bankruptcy appeared in the *Gleaner*.

BANKRUPTCY

SAMUEL LOWE & BRO.

Of St. Ann's Bay

Offers in writing will be received by the undersigned up to 4 p.m.

On Friday, 10th November 1931 for . . .

And then came a long list of his inventory that reveals the scope of his ambition and the range of his hopes and aspirations. This was not just a small dry goods store; this was a store that intended to meet every possible need of every possible customer. Was someone in need of a fortifying drink? Wines and spirits were down one aisle. Perhaps a hammer and some nails to do a few repairs in the house? A corner of the shop was a small hardware store. A kitchen could be completely stocked not only with groceries but with crockery and pots and pans.

Someone who was feeling poorly could perhaps be cured by one of the patent medicines in an impressive display case. A guilty husband or a courting lover could find some perfume or feminine stationery. Then there were tables, benches, ice chests, small showcases, and scales, including a Fairbanks scale that cost 413 pounds sterling. In addition, counters, a typewriter, an iron chest, a desk and chair, his precious electric generator, furniture, and household utensils were all for sale, and I wonder if some of these items were not from their living space rather than their retail space.

Reading this list of items makes me feel as guilty and unhappy as looking at a collection of bedraggled, abandoned belongings in a front yard after a foreclosure. All the aspirations are there, and the sheer hopelessness of the circumstances makes them pathetic.

Not only was Samuel bankrupt, but Swee Yin was pregnant with my aunt Barbara Hyacinth Lowe and he still had to send money to China for his other children. His only option was to go to Kingston and attempt to do something else. Samuel still had his car, and he began a modest business delivering goods to small shops. But the noose was tightening around his neck. By 1933, another baby, Anita Maria, was born; he now had two small children. Meanwhile his prospects for making money had grown even slimmer. What tormented conversations must have taken place between him and Swee Yin before they decided to return to China and to the rest of their family?

On July 3, 1933, Samuel Lowe, his wife, and their two daughters boarded the ship *Adrastus* to return to China. Along with them were many other Hakka Chinese who had resided in Jamaica and were now facing the reality that the social and economic situation there was too dire for them to remain. The journey home started in the capital city, Kingston. Samuel and his family stood on the deck; filled with longing, he gazed at the shore as the ship slowly left the "island of forest and water." He had gone back and forth to China four times since he first arrived in 1905, but as the *Adrastus* sailed farther into the southwestern Caribbean, and he looked at the receding coastline of Jamaica, he must have known that he would never return.

My grandfather had spent over twenty-five years in Jamaica, from the time he arrived on the island as a teenager until his departure. They were years of hard work, adventures, happy and ultimately unhappy twists of fate. And yet, he always considered them his best years. He loved the island and its opportunities, the women who had shared his life, the children he had fathered. He must have been overwhelmed by his emotions. There is an old Hakka saying: Birds that have left their nests will always remember their home; the drifting boat will eventually desire to return to shore; and busy travelers will forever yearn for their hometown.

YOU DEN, YOU CHOY: HAVE CHILDREN, HAVE WEALTH

S UNHAPPY AS HE MAY HAVE BEEN ABOUT LEAVING JAMAICA, MY GRANDFAther had something to look forward to in China. After four tumultuous years, he would be reunited with his mother, and the four children whom he had left earlier and who were waiting for their parents' return. He and his wife had been terribly worried about my uncle Chow Woo. After nearly two years of living in the village, where sanitation was, at the time, rudimentary, Chow Woo became very ill; eventually he was passing blood in his stool. When the news of his illness reached Jamaica, his mother, now nursing yet another baby, must have been frantic. She wrote to her family in Guangzhou to find out if she could send the children to them from Lowe Shui Hap to be kept safe, and for Chow Woo to recover—her father's herbal remedies might restore his health. Her parents agreed, and the siblings were reunited at their maternal grandparents' home in Guangzhou in 1930.

Finally my grandfather and his family of four, including a toddler and an infant arrived in Hong Kong in the autumn of 1933.

Their first stop after returning to China was to visit the ancestral village of Lowe Shui Hap, where Samuel had grown up and his mother lived. His mother, whose name was Ye, was delighted to see

her successful, adventurous son and her daughter-in-law. Before
they arrived, Swee Yin had dressed the grandchildren in their finest,
most formal outfits for their presentation to the elders of the family.
No longer itinerant, Samuel and his family now began their lives as
Hakka in China. The moment they arrived their first obligation was
to pay homage to their ancestors—to *our* ancestors—and visit the
bones in the village temple. These were the same ancestral relics that
I would later be introduced to by my uncle Chow Woo, the same
graves at which for hundreds of years our ancestors, holding burning
incense, had bowed in prayer.

After this ceremony honoring the dead, it was time to honor the
living: friends and family. My grandfather gave a huge banquet to
express his appreciation to them for all their help during his long
absence.

The return home must have given Samuel some new energy,
some new ambition to repair his recent losses with future profits. He
returned to Guangzhou, where he was reunited with Swee Yin's family and his four older children. Again, the first priority was to meet
and show respect to the elders, his in-laws, who had cared for his
children.

At last he could see, embrace, speak to, and admire Adassa, Chow
Ying, Chow Woo, and Chow Kong. How tall, healthy, and thoroughly Hakka they had become! It was more than four years since he
had seen them, and they had grown and changed astonishingly. How
sweet and serene fifteen-year-old Adassa was! How was it possible
that she was now nearly a woman? His oldest boy was already ten!
Samuel looked at his large, handsome family, healthy and reunited.
The older siblings crowded around their younger sisters. Perhaps
Swee Yin wept. Perhaps Samuel did as well. Perhaps leaving Jamaica
was not so bad after all.

He had hatched a plan during the long days and nights of his
journey home. His first stop was at He College in the Yuexiu district
of Guangzhou. This college was a home base for students with the

surname He from the Hakka Jiaying state in eastern Canton. During the Qing dynasty, the He students would stay there to prepare for the imperial examinations held in Guangzhou. In fact, the Hakka architectural features of the college are still clearly visible. It was turned into a residence for the traveling Hes who came to Guangzhou on business after 1905, when the modern education system replaced the imperial examinations. One of them was Swee Yin's grandfather Ho Chit Sang.

Even though they still owned two properties in Guangzhou, the couple decided to stay in He College for completely practical reasons. For one thing, they needed the income from renting the tenement to the printing company. Second, although the space was sufficiently large and versatile to accommodate residences or industry, in its present state it was uninhabitable for his large family, so a printing business was a perfect tenant. Third, Guangzhou was the home of Swee Yin's family, and my grandfather may have felt that he owed her this much after all the years of sacrifice and travel she had endured.

Swee Yin also felt guilty about the burden she had placed on her family by leaving her children with them. For the ten years she had been married, she had not had a chance to fulfill her filial responsibilities, and these responsibilities seemed even more important now that her parents were older. It was her turn to stay and look after them, even help to support them financially.

Again, the best intentions were thwarted. The plan to stay together was simply not feasible. There were eight people in Samuel and Swee Yin's nuclear family, and another ten in Swee Yin's family. The cost and logistics of feeding everyone every day became overwhelming. Her parents and uncles and aunts loved the children, but grandparents throughout the world—and I am one of them—can appreciate that, much as we love our grandchildren, we have also come to love our times of silence and peace.

Samuel and his father-in-law immediately began renovations on another building on Chaoguan Street. It was an old-fashioned tenement with two and a half stories, directly across the street from Swee Yin's parents. The solution was perfect, giving Swee Yin easy access to her parents and siblings while also giving her own sprawling family the autonomy that was essential.

My grandfather's drive and ambition once more took effect. Real estate was the way to go now that he had returned to Guangzhou, and all his saving and his sending money home over the years began to pay off. He decided to transform the run-down tenement into a three-story mixed-used building. The first level was for personal use and also served as a shop front for his business; the second and third floors were to be rented out to students from the nearby Guangzhou Second Middle School (which still exists today). The renovation took about half a year, from the second half of 1934 to the beginning of 1935.

Samuel moved his family to the new space, but after living there for only half a year, Swee Yin found even this short distance from her parents intolerable. Again the family moved back to 29 Chaoguan Street. They never returned to the beautifully renovated building, and Samuel decided that his two nephews should manage one of the shop fronts. The second and third floor continued to provide my grandfather with rental income.

Now that he had found a comfortable residence for his family, he immediately started to look into business opportunities. My grandfather was emboldened by his world travels. He was self-confident, knowing that since he had succeeded in a foreign country—and if not for the fire and the Depression, his success would have been even greater—making a living at home would not be a problem.

I am always struck my grandfather's business acumen and imagination. When I look at my own business sense and Elrick's savvy, I wonder if there is something in our DNA that facilitates making money, and if it came from our grandfather. The streets of Guangzhou were bustling with enough people to support many small businesses—and the university life would provide still more customers: the students had a little money to spend and were in search of some new experiences. And everyone was hungry at some point during the day.

A restaurant. Samuel would open a restaurant.

He had always been an exceptional cook, and the locals in Guangzhou had a great appreciation for food. But where? What would be the best place for his new venture? The busy Zhonghua Middle Road, less than half a mile long, was a perfect spot, with plenty of foot traffic. It was a brand-new thoroughfare, built in 1930, and crowned by the impressive "Four Arches" buildings. This commercial road was highly specialized; on its northwest side, electronic components were sold; the Tao Street part was the "electronics street," with small home appliances, cheap radios, and record players; and you could purchase clothing around the corner from the "Four Arches."

Samuel's new restaurant, Bie You Tian—"Heaven on Earth"— had its grand opening a year after he arrived home, at the beginning of 1934. My family, like all Hakka Chinese, is especially sensitive to historical precedents, to the continuity of generations and achievements. Family members often refer to the Honorable Luo Ruifeng, who was the founder of the New Crane Lake Lowe clan. He was forty-four years old when he moved to Longgang and established his empire. The story of how the Honorable Luo Ruifeng made his first bucket of gold goes like this:

He noticed that people who were selling produce in the market at the time always carried some Hakka wine with them in small flasks. He had a taste of it and concluded that there was a market for it. He also realized that his own home-brewed wine was far better. He then decided to abandon his agricultural business and focus on producing wine. He had acute market sensitivity, and was able to grasp opportunities as they showed up; that was the key to his success in his business career.

Put simply, he made a fortune.

He was also lucky. The time, the place, the supply, and the demand were all in perfect harmony.

Our grandfather noticed a parallel in his own story and saw it as a good omen. He too was forty-four years old when he moved to Guangzhou to reestablish his business in 1933. Perhaps this coincidence would bring him the same kind of luck. The impressive precedent of his ancestor was always in his mind as he embarked on this new chapter of his life.

Timing, as we all know, is everything. And Samuel Lowe, like many others during these years of the troubled, dramatic twentieth century, had a difficult challenge. Luo Ruifeng started his business at the peak of the Qing dynasty, when the possibilities seemed limitless and the stars all seemed aligned for success. Our grandfather, on the other hand, lived in a time of turmoil rather than stability; economic recession, indeed a worldwide Depression, rather than an economic boom; and worst of all, continual warfare rather than enduring and stable peace.

If there was a certain naïveté in my grandfather's attempt to enter a new and highly competitive business during a period of great instability, there was also a certain grandeur in his insistence on following his dreams, trying something different, making something out of nothing.

There was, however, another serious problem for him. The commercial dynamics in China were changing rapidly, and he had a terrible time figuring out how to adapt to them. Guangzhou at the time had already established a thriving restaurant business. Throughout this region of China, Guangzhou was known for the quality and variety of its food. Restaurants, tearooms, teahouses, and snack shops were everywhere. "Heaven on Earth" was a beautiful name, but somewhat grandiose for a rather modest restaurant serving Guangzhou locals. Samuel must have had visions of his crowded Jamaican storefront, with people sitting around and eating and drinking together.

Such was not to be in the bare-knuckled competitive world of

Guangzhou. To stand out among its competitors, Heaven on Earth needed to specialize in Shun De cuisine. Shun De, a beautiful region of Guangzhou Province, in the Pearl River delta, has some of the most sophisticated Cantonese cooking in the area. Because it originated in a delta, delicate fish specialties are one of its hallmarks. Cantonese cuisine tends to be more straightforward—some would even call it bland. But Shun De focuses on rich flavors and the use of fruits like dates and citrus, even milk products, combined with more traditional ingredients.

But my grandfather did not offer a good variety of dim sum during morning tea and lunch, and the restaurant had nothing distinct to offer the more sophisticated palates of Guangzhou at dinner. Heaven on Earth was closed in less than a year.

This failure taught my grandfather that he had to bring his business skills and knowledge up to date, instead of drawing solely on his experiences in Jamaica. Competing in Guangzhou might have been too demanding. But the city was surrounded by smaller villages, and these might provide a much less competitive, more self-contained market. In the first half of 1935, Samuel stumbled on an unlikely opportunity through his brother-in-law: the two bid for, and eventually won, a contract to operate an abattoir—a cattle slaughterhouse—in the town of Dan Shui in Waiyoung County, Guangzhou Province.

Dan Shui had a long, illustrious history as a market town. At the end of the Song dynasty, it had a market fair called Shang Xu ("Up Fair"), a name that was later changed to Guo Du Zhen. During the Ming dynasty (1368–1645), Fort Dan Shui was built to guard the Da Ya Wan coastline. The combination of sea and land transportation made the town a perfect location for economic activity, and it expanded and became a magnet for business.

After the Opium War in 1839, it again became a hub of trade and transportation between Waiyoung County and Hong Kong. There were so many commercial entities that streets were built for specific industries—"Big Fish Street," "Walking Pig Street," "Lantern Street," "Rice Street." As long as he could find reliable helpers, Samuel Lowe was fairly confident about venturing into the Hakka community. There is a Chinese saying: "Don't go hunting tigers without blood brothers; and don't let a father go into battles without training his sons." Grandfather met with his cousins in Lowe Shui Hap and brought them to the slaughterhouse after negotiating terms and conditions.

Superficially, a slaughterhouse might appear to be a straightforward matter of supply and demand, buying and selling. In fact, it was a multitiered enterprise that required a number of highly skilled professionals. Our grandfather knew the Hakka people very well and was able to recruit a team of butchers, buyers, wholesalers, and retailers who worked well with each other.

The system of slaughtering livestock at the time went something like this: Cattle traders had to go to one of the slaughterhouses and pay a fee based on a standard price chart. The owner of the slaughterhouses then had to pay a flat tax to the government every month on a specific date. Cattle were mostly used as draft animals in agriculture, but as the economy grew, they were increasingly needed for meat. Grandfather's business was now growing at a steady pace; rental incomes were flowing in every month; and his children were growing up healthy.

Finally he had good reason to be very hopeful about the future.

And then came the catastrophe of the Sino-Japanese war.

THE RAVAGES OF WAR

TENSION BETWEEN CHINA AND JAPAN HAD EXISTED SINCE THE NINETEENTH century, long before the second Sino-Japanese War began on July 7, 1937. China was first defeated by Japan in 1895, after Japan infringed on Chinese sovereignty and territorial integrity by seizing control over Korea. Korea had long been an important satellite of China, but with its rich natural resources and its strategic location close to Japan, it was also too tempting to ignore.

An internal rebellion in Korea attracted both Chinese and Japanese troops dispatched to quash it. Political maneuvering left Japan stronger and China weaker. Korea, encouraged by Japan, declared its independence from China, which battled but lost to the more modern Japanese army. The 1895 treaty granted Korea's independence, gave Taiwan and other territories the right to secede from China, and levied a huge fine on China.

Over the next twenty years, China defeated Russia to gain dominance over Manchuria, which was a part of China until 1931, when a battle known as the "Manchurian Incident" left it under Japan's—not China's—control.

———

It was in the Marco Polo Bridge incident of July 7, 1937, when shots were exchanged between Japanese troops conducting maneuvers in north China and a local Chinese garrison, that China's biggest defeat at the hands of the Japanese in modern times occurred. The Chinese tried to settle the matter quickly by offering concessions, but the Japanese used the incident as an excuse for sending a large force to north China. Before the month was over, it had captured Peking. The Chinese government fought back, but only reluctantly, and in one major defensive effort the Chinese lost about 250,000 men.

My grandfather and his family were far enough south to escape the actual fighting thousands of miles to the north. But news from the northern provinces was catastrophic for him and many other struggling families.

The Marco Polo Bridge incident led to a full-scale invasion of China by the Japanese imperial army and to the second Sino-Japanese War. This war proved to be disastrous for my grandfather's family business. His abattoir in Dan Sui, from which he had expected high sales and employment for his relatives, had, like his real estate in Guangzhou, degenerated from a promising economic challenge to a potential liability. It was still functioning, but vulnerable.

On August 31, Guangzhou suffered its first air raid. In a panic, most of the inhabitants—my grandfather included—decided that the city, despite its vibrant commercial center, was unsafe, and there was a mass exodus. Grandfather rushed home from Dan Shui, quickly packed, and brought his family to Lowe Shui Hap. There were several apartments in the village that had been distributed to the siblings after the death of Samuel Lowe's father. These came in handy during this emergency. Samuel and his family shared a unit with the family of his elder brother Hin Chiu. Although it was crowded with children and in-laws, and a far cry from the spacious house in Guangzhou, they renovated the unit and lived there quite comfortably.

The slaughterhouse in Dan Shui was still a viable business, and my grandfather had to return there immediately after having settled his family. He left his wife with the task of arranging new schools for their children, some of whom were still quite young. Chow Woo and Chow Kong were enrolled in the Pinggang Middle School. Adassa did not attend school in Guangzhou, but Barbara and Anita Maria went to Yiyan College in the village. Chow Ying, who was fourteen at the time, had always despised school and was put to work to help out the family.

Worries about returning to Dan Shui were easily rationalized away. It had always been a peaceful spot, and it was remote from some of the dangers in Guangzhou. But soon after Samuel Lowe arrived, its peace and quiet were shattered by the war. A brigade of the Japanese army arrived onshore in the Xia Chong area on October 12, 1938. The Japanese invaders were ruthless, and they took over the slaughterhouse, leaving my grandfather no options. He was forced to close it down, though he fortunately managed to escape and return to his family in Lowe Shui Hap.

He might have reassured himself that he still had a rather valuable piece of Guangzhou real estate. But in 1939, Samuel Lowe got the devastating news that his buildings at Guan Yin Shan (Guanyin Hill) had been torn down by the Japanese army. The Japanese had decided on urban destruction in order to accommodate the transit of munitions: they tore down buildings to widen streets for the convenience of army trucks. That same year, Chow Woo and Chow Kong had their schooling interrupted again when the Pinggang Middle School stopped all its classes. Grandfather's and his family's quiet village life in Lowe Shui Hap was replaced by the sound of gunfire and bombs. From 1938 to 1945, the Japanese imperial army occupied Guangzhou and with the occupation came unspeakable horrors, including infamous medical experiments on prisoners involving deadly bacteria.

With all this, and with no reliable income, my grandfather reached the lowest point in his life. He was unable to provide for his large family, and finally he and his wife said that all the children except the two youngest girls needed to find work, for the common good.

My grandfather scraped out a small profit from the sale of goods and medicines he imported from Hong Kong. But that business became too dangerous after the Japanese captured Hong Kong. The marauding Japanese troops caught him a few times, confiscated his goods, and beat him for doing what he needed to do to keep his family from starving. Eventually, he ended up butchering livestock at the Longgang fair, an event that continued despite the war.

The entire Lowe family pitched in to keep going. Swee Yin and Aunt Adassa set up a roadside stand just outside the ancestral village gate, selling meals and tea to refugees who were fleeing the Japanese. Aunt Adassa had married a chef whose specialty was Western cuisine, and the young couple moved to Guilin in Guangxi Province, about 350 miles from Lowe Shui Hap. Her brother Chow Ying eventually joined her there and sent home money that he earned transporting passengers on a bicycle Samuel Lowe had purchased for him. There were few automobiles then, and bicycles were the preferred mode of transport in most villages. Thirteen-year-old Chow Woo was still a student in the Chinese equivalent of junior high school, but education was less important than sheer survival. He went with his uncle to a small town in Heyuan County a few hundred miles away, where he became an apprentice taxation officer.

The family's financial troubles persisted and the war raged on. The family abandoned the lowland village of Lowe Shui Hap as word spread that the Japanese were approaching, and sought safety with friends who lived in the mountains around Shenzhen. Eventually, in 1945, the Chinese defeated the Japanese, at a nearly unbelievable cost of lives. My grandfather must have thought that peace and prosperity were at last within reach. But again, history made this impossible.

In 1945, after the war with the Japanese but before the civil war that would take place in China, my grandfather returned to Guangzhou with Chow Woo, Barbara, and Anita Maria, and they all lived in their tenement at 76 Longzang Street. Japan's surrender had heightened the tension in China between the Nationalists and the Communists, whose conflict had been suspended during the war against Japan but was unleashed when it was over. Four years of internal fighting devastated China. The economy was destroyed, inflation was rampant, and Guangzhou was again in a panic.

Samuel Lowe was now almost sixty years old. Today, as I can attest, that age is a vigorous phase of middle life, but then it was the age of an old man, and Samuel was financially and spiritually ravaged. His adult children took on the responsibility of helping their parents financially, while our grandfather dabbled in a number of small business ventures.

He had passed along the gene for entrepreneurship, it seems. Aunt Adassa and her husband, Zhangpin Liu, together with Uncle Chow Ying, returned from the Sixteenth Team at the U.S. air base in Guilin and started a restaurant and a transportation business. Uncle Chow Woo and his uncle were transferred to the Qingyuan branch of the Sanshui taxation office as clerks; Uncle Chow Kong had just started working in the Union of Privateers. So the Lowe family managed to survive, intact and in many ways modestly prosperous. The combined income of parents and children, and the rental income from two floors of the tenement, provided enough money to cover daily expenses and tuition fees for the young students, Barbara Hyacinth and Anita Maria.

The family may have suffered great financial losses, but the family bonds had become stronger than ever.

THE GREAT PROLETARIAN
CULTURAL REVOLUTION

WHEN THE PEOPLE'S REPUBLIC OF CHINA WAS FOUNDED IN 1949, SAMUEL Lowe was officially retired. This should have been the time, I imagine, when my many first cousins heard our family's wisdom and history from our grandfather. But they remember very little of such conversations and in fact, few ever happened. The Communist regime affected even the most intimate family relations.

In the early years of his retirement, Samuel Lowe did not have a tranquil life. The government, suspicious of any germ of capitalism in the new Communist society—for such a germ was seen as potentially infecting the body politic—began to scrutinize everyone who might have had a life elsewhere. My grandfather never made a secret of his time in Jamaica, and he was immediately targeted as being politically suspect.

One problem was that he had once been a member of the Zhi Gong Tang, an organization formed to unite and protect the Chinese who lived overseas. However innocent—indeed, however patriotic—this group was, it became an object of suspicion and contention. My grandfather was put under intense investigation and was only partially exonerated when the investigators discovered that what he had told them from the beginning was true: the organization had no ambitions in mainland China. Even though he was cleared of the charges, he could not rewrite his life story; and because he had lived in Jamaica there was a cloud not only over him but over his children. My aunts' and uncles' careers and marriages always had the slight taint of their father's previous life, and of their Jamaican passports.

Small wonder that he simply didn't want to discuss it.

The Communist Party was intent on categorizing what it referred to as the "family composition," which was a way of insinuating the bureaucracy into the most private and personal aspects of a citizen's life. Its impact was so great that it changed the fate of many Chinese families, and here too my grandfather could not dodge the investigation.

The issue concerned family-owned properties that were eventually, forcibly nationalized. Even though they were landowners, my grandfather and his elder brother received only a small vegetable patch in 1948 when Lowe Shui Hap was evaluated and its village land was distributed. My grandfather and his brother were immediately transformed from the status of "landlords" to that of "poor peasants," while many of those who received a larger allocation of land were categorized as "landlords." The demotion proved to be a blessing. A large number of those designated as "landlords" were immediately targeted politically, and were unable to escape oppression for the rest of their lives.

The Great Leap Forward, Mao's tragic and demented scheme to transform China from an agrarian to an industrial economy, began in 1958 and lasted until 1961. Chinese peasants' land was collectivized and the peasants themselves were placed in industrial settings. The result was a famine of almost unimaginable dimensions: the estimated death toll is at least at twenty million Chinese, and that figure does not include the lurid murders that also probably took place. The Chinese political elite was divided and Mao's powers were reduced. His radical, failed policies were modified as planners tried to stabilize the economy, but Mao felt threatened and mobilized his followers to reinstate him as the essential, irreplaceable leader of the revolution.

By 1963, the "little red book," *Quotations from Chairman Mao*, was studied widely, especially in the People's Liberation Army, where he was still revered. Mao was unhappy about the course of the "revolution," about other leaders "taking the capitalist road." And, of course, he was angered by what he considered (rightly) to be efforts to shunt him aside. He set about tearing Chinese society apart to bring it closer to his vision, and the place to start was the Chinese Communist Party hierarchy: he purged its leadership and many midlevel officials, and chaos resulted throughout the country.

Young Chinese, mostly as Red Guards, spurred on by Mao and his allies, attacked their parents and their teachers. Party officials encouraged the Guards to beat thousands to death and to drive thousands more to suicide. Targets of the rampage included intellectuals, artists, and anyone or anything even remotely related to the West. My grandfather, of course, was an obvious target. But he managed to keep a low profile until the Great Proletarian Cultural Revolution was launched in May 1966.

This was Mao's most audacious and terrifying crusade. It lasted roughly two years. By way of solidifying his power and using mass terror to ensure national obedience, he decided that all possibly unreliable social and political actors in China had to be purged, so that a purely Marxist Communist regime could finally take hold in the vast country. In the purges that ensued from 1966 to 1968, so many millions of people were persecuted that the leaders could not rely on the Communist Party alone to do the job. They enlisted the "masses" or, to put it more bluntly, the mob to find and destroy class enemies—businessmen, the educated elite. This was done through various techniques: murder, imprisonment, harassment, theft, the destruction of relics and precious Chinese artifacts. But it all boiled down to a reign of terror.

My grandfather was a confirmed capitalist, but the years under

Communism had worn him down. He was quite old by then, and he suffered throughout his last year during the revolution. When the mob took over Guangzhou—which had been renamed and no longer was called Canton—in the fall of 1967, conditions were so chaotic that the important Canton international trade fair had to be postponed for a month. In early November 1966, several hundred thousand Red Guards from other parts of China had streamed into Canton, where they were joined by local students. Their role was to overwhelm the city, leaving no home unsearched, no individual unscathed by the blaze of revolutionary fervor.

By mid-December 1966, rival Red Guard factions were fighting each other in the city streets, until finally, in January 1967, party officials surrendered their offices to an alliance of rebel organizations. Leading party officials were "placed in detention," and the main struggle was no longer between them and the Maoists, but between different coalitions of the Maoist Red Guards. This was not what Mao had planned, so the army was called in to maintain order and support the revolutionaries. Military control of the city was established officially on March 15, 1967, and in April the precious Canton trade fair scheduled for the spring was opened by Vice President Zhou Enlai in person.

In this chaos my grandfather died. He was already ill in 1967 and while he was not important enough to be targeted specifically, he must have experienced some despair and worry over what had been taking place. Uncle Chow Woo's company held a large, formal funeral for him, attended by a crowd of people he had known, and friends of his many children. The funeral must have been a rare peaceful moment during the most violent year since the Communists had come to power. Local fighting intensified around Guangzhou through mid-August. Rail transport was disrupted; food shortages affected all the citizens; buildings were gutted by fire. People had to walk long distances to work and were fearful of venturing out at night, when hoodlums terrorized passersby and one another. Fights broke out in previously quiet neighborhoods, and the violence amounted to anarchy, beyond the control of the police. Student Red Guard factions brandished guns and fought at the university. Many thousands of people were injured and hundreds were killed throughout the city.

By 1967, the discontent of the peasants in the provinces escalated as they sought relief from starvation and oppression. They flocked to the cities, which were ill-prepared for any further social dislocation. There was mass unemployment in the Communist paradise, and urban people who did have jobs endured horribly unfavorable working conditions. There was a reverse migration: while peasants flooded into the cities, unhappy university students were deported into the countryside when their schools closed. The Korean War veterans, who had endured their own share of terrible suffering on the front, returned home to even greater deprivation, lacking any decent skills for civilian work.

For four years, beginning in the summer of 1966 and not petering out until 1970, this cataclysm dominated the Chinese experience. What was invisible to everyone was a struggle to the death within the Chinese leadership, on determining the course of the future. While Mao lived the excesses were conveniently blamed on others, but after he died in 1976, the so-called Gang of Four—led by his wife—was arrested and charged with treason.

All this was after my grandfather died, of course. He was seventy-eight years old at his death, and he certainly had been ill, but the stress of the political revolution could not have helped. I asked one of my cousins for any documents—letters Samuel had written home from Jamaica, bills of sale, anything in his handwriting. What about the anecdotes that our grandfather might have shared? Impossible, the cousin replied. "Our grandfather was branded a capitalist." The family destroyed all of his records, which would have been seen as incriminating evidence of his capitalism. "The government took away his houses and his money. He knew better than to share nostalgic tales of his life as a businessman in the West."

When he was a teenager, my grandfather boarded a boat that was headed to an unknown country. He intended to take care of himself and his family and send money home. Then his brothers came and he built a business and collected as many of his children as he could and finally returned to China, with all its turbulence and tragedy.

The final years of his life should have been peaceful. But they were years of turmoil, of a reign of terror, of genteel poverty. The more I learned about it all, the more I understood. Yes. Of course he died from a long illness, as many seventy-eight-year-old men do. But I felt strongly that the political and social foment of the times also broke him. "I know that is what killed you," I said to him quietly. "They broke your spirit and your heart."

How curious that, for me, 1967 was a time when my heart and my spirit came close to breaking as well.

AND THE FAMILY
GROWS AND PROSPERS

UNCLE CHOW WOO, AUNT ADASSA, AND THE REST OF THEIR FAMILY HAD TO carry on in a China that had been radically transformed and traumatized, just as they had been radically transformed and traumatized by the death of Samuel Lowe. They may have felt some tentative peace, but they continued to live in fear, awaiting the next swing of the pendulum. I have tried to probe their memory of those times, but reticence and painstaking selection of words had been a fact of their lives for decades. To look at their serene faces, to see the large and successful families that they created, makes it difficult to imagine the dire circumstances under which they struggled.

After my grandfather died, the Cultural Revolution waned, as the army gradually brought the Red Guards under control, broke up their bands, and sent them back to wherever they came from. Terror, which had been greatest in cities like Guangzhou, subsided. In fact, people who traveled to Guangzhou at the time said that it had a different feel from the rest of China, especially Beijing. The inhabitants of Guangzhou seemed less afraid; there were more televisions; and the Friendship Stores where foreigners shopped were filled with Hong Kong Chinese buying things like electric rice cookers for relatives and probably for themselves, on the assumption that Chinese goods were cheaper. When we visited to celebrate the 2014 Lunar New Year, a cousin from Indianapolis noted that the Friendship Store was no longer the go-to place for bargain hunters. In less than ten years, an explosion of other retail outlets had made these pillars of Communist commerce quaint relics of a bygone era.

Many bureaucrats, intellectuals, students, journalists, teachers, diplomats, and others who had been accused of being rightists during the Cultural Revolution began to return from the countryside. Many of them were gaunt and fearful, with the haunted look of people who have been abused. Indeed they had been badly treated; crowded into camps, doing manual labor, living in pigsties, and working on farms and in factories under arduous conditions. One university president spent those years as a janitor.

Deng Xiaoping and his cohort were brought back to Beijing in the mid-1970s, but few were secure until after Mao's death. It was probably 1980 before most people were able to resume what passed for normal lives—and even then, many still lived in fear that the party hacks who continued to run numerous organizations would turn on them again. To a greater or lesser extent, they had lived in terror for nearly twenty years, and that experience changes people and changes society. Before the late 1970s, the peasants were the only ones who were relatively spared. As the thaw began, they might have been glad to be rid of some incompetent workers—those soft intellectuals and city folks who didn't know what to do with a shovel, much less a farm animal, but had been sent to the countryside to work and be reeducated. But for the most part, the Red Guards had left the peasants alone.

Invisible to most eyes were fierce power struggles at all levels of government, from national to local. Who would win: the radicals or the moderates? And what did those terms mean after Mao?

Slowly, in the 1980s, reforms were introduced. The peasants were the first to experience major gains as Deng launched "household responsibility," allowing them to work on private, individual family plots. While many farms were still collectivized, peasants also could cultivate their own piece of earth. Agricultural production soared, and many peasants became relatively rich.

That was only the beginning; eventually, major economic reforms allowed China to become an important participant in world trade, and to obtain large foreign loans—all of which enabled ordinary people to improve their lives significantly. My family took advantage of every opportunity. A Hakka, Deng could have been echoing my mother when he said, "To be rich is glorious." (He also said, "Poverty is not socialism.") What a shame that my grandfather was not alive to hear it! This was a time of flowering in our family. All the pent-up entrepreneurial energy was released.

My uncle Chow Woo and two of his children worked as accountants in a large seafood distribution company. Later, one of the sons started his own seafood company, which became a family business, employing some young people from the next generation. His son is now an executive in this family-owned business. Aunt Adassa's son moved to Hong Kong, where he owns a large hardware company. Aunt Barbara's children went into the import-export business and are now consultants to companies in China, Great Britain, and, fittingly, Jamaica. Following the family's commitment to education, one of Aunt Anita's sons owns a for-profit university system in Australia. Then there are those in government service: a daughter who is a judge in Guangzhou and a son who is head of the antinarcotics division of customs for Guangdong Province.

The Lowe family lived in the right part of China. Guangzhou moved faster than much of the rest of the country in becoming prosperous. Hong Kong was close and Beijing was far away. Shenzhen, which was next door, was designated China's first Special Economic Zone, and this designation drew billions in foreign investment. But there were paradoxes. On the one hand there was a flowering of economic opportunity. On the other hand, political repression was still severe. There was no interest in importing Western values such as free speech and democratic elections.

In the 1980s, there was a growing sense among interested students and intellectuals that China would soon open up politically, but they were wrong. In the mid-1980s Deng initiated two movements against "bourgeois liberalization" and foreign values. In the 1990s and ever since, the rich have gotten richer; rural income has stagnated; and peasants have fled to the cities, especially Shenzhen and Guangzhou, to work in factories. Many villagers, exploited by corrupt local officials, have migrated illegally to cities where they have horrific lives.

China also became wide open to foreign investment and here too my family—my sprawling, talented, entrepreneurial Chinese family—was involved. Entrepreneurs like my cousin were permitted, even encouraged to open their own businesses; and in the late 1980s and 1990s many state industries were privatized and their price controls were lifted. And, in the spirit of the great circle of history and of life, the Chinese have returned to Jamaica, not as indentured workers but as partners and entrepreneurs. Three of my aunt Barbara's sons have lived in Jamaica, from six months to two years, and established a global import-export business. Aunt Barbara was born in Jamaica, and she returned there to visit her sons and reconnect with her birth country.

I was struck by a photograph of the prime minister of Jamaica, the elegant and impressive Portia Simpson Miller, shaking the hand of Xi Jinping, president of the People's Republic of China, when she visited Beijing in September 2013. In some ways it was just another state visit, just another "grip and grin" photo opportunity for two leaders—one from the world's largest country and the other from one of the smallest. And it made sense that they would have a formal meeting, given the fact that trade between China and Jamaica had tripled between 2009 and 2012.

But in other ways this was an emotional and dramatic image for me, the daughter of Nell Vera Lowe. The occasion of a proud Black Jamaican female head of state shaking the hand of her Chinese male counterpart was something that my mother would probably never have predicted. Indeed, it was something that Samuel Lowe could hardly have imagined. It represents a momentous change, but there are many elements that are familiar. The Chinese are coming to Jamaica from a position of economic strength and dominance. This time, mixed in with the Chinese workers on the sugar plantations that were purchased by the Chinese, there are also white-collar professionals. And there is resentment among the people of Jamaica about the Chinese influx, similar to what my grandfather experienced.

These two countries seem mysteriously destined for each other—so passionately divided, so improbably connected. In the streets of Kingston new shops owned by Chinese have opened. And I have to wonder if my mother's and our family's narrative is timeless. Will another Chinese man become involved with an African-Jamaican woman? Will they have a daughter together? Will he decide that he needs to marry a Chinese woman? Will his daughter then lose her father? Of course I will never know. But this is no longer just the story of my mother and her father—it is an important part of the sweeping history of two worlds, two countries, and countless lives.

PART SEVEN

MY THREE-THOUSAND-YEAR-OLD FAMILY: REUNITED AT LAST

Why not help each other on the way?
Make it much easier.

—BOB MARLEY, "POSITIVE VIBRATION"

VISITING JAMAICA

IN OCTOBER 2012, BOTH OF MY BROTHERS; MY FATHER'S BROTHER, UNCLE HARRY; and Keith Lowe, our cousin who now lives in Toronto, took a trip to Jamaica together. Our ostensible reason was to attend a Caribbean economic development conference sponsored by the *Carib News*. That year, Montego Bay, Jamaica, was the site and I was asked to appear on a panel to discuss the entertainment business.

The reason might have been the conference, but my brothers and I were restless, and the discovery of our family in China had intensified our need to assemble other pieces of our lives and our parents' lives. We had all visited Jamaica many times, but never before did we have such specific details of our grandfather's life, and never before was our search for the places where he lived, and our curiosity about how he lived, inspired by such a vivid sense of his actual biography.

Never having spent much time in Montego Bay, which is largely a place for tourists, my brothers, our uncle, and I left there as soon as the conference ended. Our first destination was Mocho, the town where my father's mother was born, where Samuel Lowe had a shop and lived with Emma Allison, and where many of Gilbert's relatives still remained.

As we drove away from Montego Bay, I sat in the back of the van,

watching the landscape change. We went from a lush, more popu-
lated area to one with scrubby trees and bushes and brightly colored
but more primitive houses. Gradually the houses were spaced farther
and farther apart. We sped past ramshackle homes, small eateries,
random stores, men lounging in the shade of deep awnings, groups
of children playing while mothers holding babies stood and talked.
Occasionally our uncle Harry would impart some stray observation
about the family or the neighborhood.

With my memories of China still fresh, I had a new sense of what
the word "integration" can mean. Like most Black people of my generation and my parents' generation, I grew up defined in fundamental
ways by the political and social urgency of "integration," a process of
justice involving equal access to education, jobs, economic opportunity, and legislative representation. As an adolescent and throughout
my professional life, I refused to be defined by slavery even though I
could see how slavery was used to define my people.

But integration is also a psychological process of consolidating
often dissonant experiences and bringing them into harmony. Yes,
we are all God's children, but our differences define us and alienate
us. As I looked at the landscape of Jamaica on the road to Mocho and
listened to my uncle Harry contentedly chatting in the backseat, my
mind wandered to China, where I had sat in the backseat of another
van, next to my uncle Chow Woo. When I looked at the Chinese
landscape I saw dramatic economic growth and a gleaming super-
highway threading through one of China's most prosperous regions;
and as we headed toward our family's ancient village of Lowe Shui
Hap, my uncle Chow Woo serenely pointed out places of note.

These were disparate images from worlds that were alien to each
other. But for a moment, I could glimpse in a new way what my
grandfather must have felt as he *integrated* these worlds in his own
life. He never became less Chinese—just as I will never become less
Black or less American—but in fathering children with Jamaican
women, in living in Jamaica and developing businesses there, he

consolidated realms that would never have naturally touched each other. Just as I have.

We arrived at Mocho, and Gilbert's eldest son, Tony, now middle-aged and careworn, greeted us. Emma Allison was Gilbert's mother and she had worked in Samuel Lowe's shop. My grandfather had built her a house diagonally across the street and up the hill from his store, a little lean-to that is still a shop today, owned by Jamaicans. I was happy to see Tony, but even though we shared a grandfather, he was not as thoroughly a part of me as were my Chinese relatives.

As we sat and socialized in Emma Allison's old home, Tony was astonished to learn about his grandfather. Gilbert, like my mother, never spoke much about his father or his Chinese background. He never shared the fact that he had a sister living in China or his assumption that he too was destined to go there and be reunited with his father. He had a close relationship with one of Samuel's brothers, and the sense that this relationship should somehow make his father accessible must have added to his frustration and loneliness. As a teen, he must have lived with the expectation that he would be leaving soon, going to China soon, seeing his father again—soon. But "soon" never happened, because Samuel Lowe's brother who was responsible for bringing him to China died.

Gilbert was defeated by fate. He married a Jamaican woman and had ten children. But he would sit on his front porch, cross his legs, stare, and smoke. I thought of my mother, who would stand at a window or on our front stoop and stare out, as if waiting for someone to appear from around the corner, or from across the globe.

Their father. My mother. The same thousand-mile stare. The same loneliness. The same depression.

An elderly next-door neighbor stopped by to talk with us. He must have been close to ninety, and he remembered my grandfather—he was the only person we met in Jamaica who had an actual memory of Samuel Lowe. I listened to him describe my grandfather as a tall Chinese man who owned the Chiney shop diagonally across the street.

I listened to him describe my grandfather as the man who built this sturdy house at the top of the hill for Emma and her children, and as I listened the tapestry of our story seemed to become richer and more varied. This neighbor must have been just a child, watching the exotic rich man who lived next door. Some memories never fade, but they may never have an opportunity for expression. At long last, because of our presence, the neighbor was able to recount a story that no one else had ever been interested in.

Still in the master bedroom at the front of the one-story house on top of the hill in Mocho is the bed in which my aunt Adassa and uncle Gilbert were born—and probably conceived. Almost as an afterthought, one of my cousins offered to show us her grandmother's bed. We went into the comfortable front bedroom and there stood a modest but sturdy mahogany bed, about the size of a modern double bed. Emma Allison would tell her other children that her daughter in China and her son Gilbert were born there. Both had Samuel's last name, Lowe. Samuel Lowe left the house to Emma, and she left it to her children and great-grandchildren. Yet none of his children or his grandchildren ended up living in the house he built.

The bed began to take on tremendous significance in my mind. I envisioned Lowe Shui Hap, the village that commemorated our family's history and importance. Just as my mother's and Gilbert's names were missing from the family tree there, so too was any indication of the larger world that Samuel had explored and in which he had lived. Wouldn't Lowe Shui Hap be the perfect place to include the history of Samuel Lowe as an example of the tens of thousands of Hakka laborers who helped build the merchant class and strengthen the economy of Jamaica and other Caribbean nations such as Trinidad and Tobago, Guyana, Panama, Suriname, and Cuba?

The history of the Hakka in China is told quite thoroughly not only in Lowe Shui Hap but in Chinese history books. What is not told very often is why there are hundreds of thousands of African-Caribbean-Chinese throughout North, Central, and South America.

Many of us never knew our grandfathers, because many of them never returned to China, and while they were fathering children with indigenous women, they never talked much about their ancestral villages. For many of us, our family's history began in the new country.

I was determined to send that bed back to our grandfather's village, and to create a replica of his Jamaican home. I was determined to tell our story and this larger story for the hundreds of thousands of Chinese who come to visit the village. The integration of worlds was not just an internal process; it could be external as well.

Tony took us to a small graveyard that was the front lawn of the house where Gilbert had grown up. This was no formal cemetery but part of a neighborhood where death was never very far away from life. The five graves, though old and neglected, were commingled in the neighbors' lives. Some of the gravestones were worn and overgrown, but on a broken headstone I could make out the name of Marianne Lowe, who died in 1923. Could she have been still another of my grandfather's children?

SAMUEL'S SHOP

THE NEXT DAY WE CONTINUED OUR JOURNEY, GOING TO SAINT ANN'S BAY, where our grandfather's largest shop had been, downtown at a prime retail location. The streets were busy, with cars, various Chinese shops, and other stores that were doing a brisk business. The shop that had been our grandfather's had most recently been a fabric store, but the proprietor had vacated the premises only a few days before we arrived.

We all walked into the shop where Samuel Lowe lived and earned his living. I was astonished at its size: this was not the kind of ramshackle little Chiney shop that we had seen nestled in the hills and small towns of the countryside; this was a vast warehouse. The walls were green but marbled by water damage, and the ceiling was painted white. There was a rusted, motionless ceiling fan. Windows looked out over the main square. I walked slowly through the place, and I was able to feel my grandfather's presence as I had never before experienced it. I could imagine him greeting customers, stocking shelves, taking inventory, turning on the lights, smiling at his wife, wondering about Nell, regarding the imposing space with pride.

For the first time in my life, I knew that Samuel Lowe and I actually walked on the same floor, looked out of the same windows, put our hands on the same walls. I stood at the window where he must have stood and saw the same square that he must have seen, and I began to cry, tears of relief but also sadness. My cousin Keith Lowe, whose question to one of his relatives in China had opened up this world for me, put his arm around me as we stood together at the window looking outside. I was comforted by his presence and felt the power of that moment.

We then wandered through the business area of the shop into the living quarters. A rusted, corroded sink; small rooms; electrical sockets; bars over the windows; bright aqua walls. As old, neglected, and abandoned as everything was, one could still feel the presence of the family who had once lived there. My tears dried and I felt a shock of joy, a Christmas-morning excitement. I looked at my brothers and asked, "Did you expect it to be so big? It's amazing, isn't it?" This was more than I had ever hoped for. This was where he actually lived! It had the immediacy of Pompeii, where life's daily moments are preserved for eternity. I am stunned at the size of this building, its strength and permanence, its prime location; and I am suffused with pride.

This must have been where our nearly sixteen-year-old mother came to find her father's brothers running the shop. Nell must have gone there filled with hope. She must have walked out broken and defeated.

I too entered with fragile hopes. But I left fulfilled beyond all expectations. Another moment of healing for my mother beyond the grave, and for me for the rest of my life.

THE DECEMBER
REUNION IN CHINA

A WEEK AFTER I RETURNED FROM CHINA I RECEIVED AN E-MAIL FROM MY Chinese cousin Gary. "December 23 is Grandma Adassa's 94th birthday and there will be another Lowe family gathering in which more family members will be back in Guangzhou. My grandpa considers it a better time to introduce all of you to the Lowe family and you can meet more family members there. . . . As to the itinerary, it is OK for you to come to Shenzhen or to Guangzhou first, but I suggest you fly to Hong Kong and come to Shenzhen first; we can go to Lowe Shui Hap for a day and you can plan another one or two days for sightseeing. Then we will go back to Guangzhou together for the gathering the next day and we could visit my grandfather Samuel Lowe's grave in Guangzhou."

I had three months to plan a trip for over twenty family members with representatives not just of every generation but of every decade of life, from two-year-olds to people over sixty-five. They lived everywhere from Jamaica to Los Angeles, and everyone had crazy schedules, plans, programs, and demands. They might have thought I should be concerned about all that, but in fact I could not have cared less. I had one goal and that was to bring the family all together for Aunt Adassa's ninety-fourth birthday. The importance of this event surpassed any other, for anyone in my family. To have that clarity made what could have been a logistical nightmare easy. It was a matter of getting tickets, booking rooms, and getting my family in line.

In keeping with a pact that my brother Elrick and I made when we were very young, my job was taking care of the family. His job was making the money. Each of us had taken on a bit of the other's job—it's not as if I didn't make a fine living, and he certainly provided some wisdom to various family members at various times. But basically, my family listens to me, and because of Elrick we have the funds to make things happen quickly and easily.

Twenty-two of us met at Elrick's Chicago apartment on December 20, 2012. The juxtaposition of December and Chicago tells you what the weather was like. Our typical August family reunions in Martha's Vineyard involve bathing suits and cocktails by the pool. Now there were down jackets, boots, hats, and gloves piled up in the hallway as we all prepared for our morning flight to Shanghai. We landed Friday afternoon at around three o'clock and waited at the airport until six p.m., when we boarded our flight to Shenzhen. It was close to midnight when the bedraggled group of Williamses arrived at the grand and imposing St. Regis.

The hotel is about a hundred stories high, and the lobby, with its gracious seating areas, is on the ninety-sixth floor. I had hoped that my uncle Chow Woo would not have waited for us, but when I got off the elevator, there he was. It was then that I appreciated the beautiful reticence and formality of my uncle, while at the same time also appreciating that in officially entering the inner circle of our family, one crosses a clear boundary that separates restraint from warmth, formality from caring. The first five minutes of our arrival at our hotel in Shenzhen offered an example of this intricate emotional dance.

While other members of my family had preceded me and positioned themselves in various parts of the lobby, as we waited for my brother to check us in, my uncle and his wife sat serenely in another area. When I saw them, I walked to them and he rose and hugged me. His wife, whom I had never met, also embraced me with real warmth. I then asked him; even though we had no translator, if he knew that this was the rest of his family. I looked over and pointed to the various groups of Black people who had gathered—my daughter and her cousins with their little children struggling to cope with the time change, my husband standing with one of my brothers. My uncle shook his head and looked at me quizzically.

I was stunned to realize that on seeing this group of Black people arriving at the Shenzhen St. Regis at midnight—people who happened to be there at exactly the same time when I was supposed to arrive—he did not assume they might also be his relatives. I walked my uncle and his wife over to where the others had gathered and made the formal introduction, which went something like "This is Uncle Chow Woo and his wife." The boundary between reticence and warmth had now been officially crossed and the time for hugging, looking closely at each other, and connecting as a real family had begun.

ROOSEVELT
FINDS HIS OWN CHINA

MY HUSBAND, ROOSEVELT, HAS A BIG PERSONALITY BUT HE DOES NOT OPEN up easily. Before we left, he told me in no uncertain terms that he was spending the holidays with my extended family in China only, and that was absolute, *only* because he loves me. All of my stories about China and my aunt and my uncle were interesting to him, but interesting *only* insofar as they involved me and my immersion in the quest. He has always loved to see me working hard, excited about a project, in a groove. Since I retired, we have worked together on a few projects, like owning and managing the LA Sparks WNBA team. But this trip—the distance, the timing, the occasion—did nothing for him. Moreover, the fact that he was born on December 24 meant that he would be celebrating his birthday in China as well. Not what he had planned.

Roosevelt, who grew up in New Orleans, came from a world totally different from mine, not only because his family was conventional and intact, but because the population of his world was much more limited. When we were first married and living in New York, I might refer to someone and say, for example, that this person was Jewish. Roosevelt would ask me what on earth I meant and how I could tell. Well, look, I would say; you can often tell by the name. He was baffled. When he was a child in New Orleans there were three basic groups: Italians, whites, and Blacks. "Wait a minute," I would say. "Aren't Italians white?" Roosevelt would shake his head and laugh: "Not for us they weren't." He grew up with almost no Latinos or Asians, no appreciation that there were many shades and nationalities of white.

Obviously, over the years his world expanded and he was exposed to diversity. We were in Africa together. He accompanied me when I went to Beijing for the Olympics. When we were in Beijing, people on the street would look at this tall, well-built Black man and assume that he was one of the athletes. He was continually stopped on the street with requests for an autograph, for a comment about his sport, or, most often, for photos. But now, five years later, he was going on a very different trip, without the distraction of the Olympic games. The only distractions he could count on were food and shopping.

"I am going only because I love you," he said, as he was packing his suitcase—reluctantly.

"I know, baby," I said. "And I really, really appreciate it."

I had a feeling that he might find out he liked it, but this was not an issue that I needed to contend. There were a lot of people I could have worried about regarding this trip: my brothers and their families, children and grandchildren, my daughter and her son, Gilbert's two daughters whom I had persuaded to come all the way from Brooklyn. And that doesn't include our large Chinese family who, I had heard, were being mobilized for our arrival.

And basically, it was all on me. I was the one who was making this trip happen. I was the one who set in motion all the systems that had to fall into place. If there were any setbacks, any problems, any disastrous events, the responsibility would fall on me. Not that I thought that there would be any such issues. Roosevelt's articulation of his reluctance to make the trip didn't worry me. My greatest desire was that my aunt Adassa would live to celebrate her ninety-fourth birthday with her whole family.

And she did.

The morning after our arrival we boarded a bus that would take us to Lowe Shui Hap. This was the same journey that I had made in August with my uncle, now repeated with four new generations of Lowes. When we arrived at our village I looked around in astonishment. My family had come out of the woodwork! There were at least 150 of us. Our second cousin was the regional communist party official, and he had the clout to arrange the necessary accommodations. I suppose it is not surprising that, given how many Lowes had assembled at the family village, the museum was closed to the general public.

With so many of us poring over the virtual book, pointing out our family line, burning three sticks of incense to honor our ancestors, my mother's absence from the family tree wall seemed an even worse affront. Gilbert's two daughters Loraine and Andrea joined us, and their father's absence from the family tree caused them similar pain. Nevertheless, with representatives from Gilbert's family there, for the first time ever, all the diverse and far-flung fragments of Samuel Lowe's family were united.

During this visit I was as much an observer as a participant in the activities. My private tour with Uncle Chow Woo in August had introduced me to some of the wonders of our village. On this day I could be engaged and reverent while drinking in the experiences of others: my brother Elrick, his camera clicking away; my brother Howard moving from one daughter to another, one grandchild to another, to greet each of them.

I watched Imani lift six-year-old Idris up to wave his hand across the virtual book, turn the virtual page, and wonder at the thousands of Chinese characters, each representing the name of someone whose life had in some way contributed to our own lives. Imani was talking to Idris in a soft voice, and even though he was still confused by the time change, he was completely absorbed by being in his mother's arms and doing this magic on the screen. His dark dreadlocks cascaded around his face and every once in a while he would look up with a question.

Once more I was struck by the beauty of all the generations who surrounded me, the miracle of the continuity of our lives and our extraordinary family history. My mind wandered to an imaginary time, a time unaffected by politics, by history, by the tragedies of this vast country, by my family's tangled narrative. And I thought of what it would have been like for me to be in China with Imani when she was Idris's age—what it would have been like to take her to Lowe Shui Hap and show her the wall, as my mother stood apart and watched the two of us together.

I took another step backward in time and thought of what it would have been like for my mother to take my brothers and me to China when I was Idris's age, bring us to Lowe Shui Hap, and show us the wall, as our grandfather Samuel Lowe stood and watched us. I thought of what it would have been like for my mother to be in Samuel Lowe's arms, as he lifted up his oldest child and showed her the wall of her ancestors in their family village. I could somehow see all of this as I looked at the wall, though I could not read the names. I saw how each of us could have owned our history in a different way had we grown up with it as Idris was going to grow up with it, rather than acquiring it so late in life. And I wondered, if we had, what difference it might have made in who we were and in how we saw the world.

Then my mind ventured into a not-so-distant future, and I imagined Idris with a child of his own. I imagined my young, beautiful Imani, not as the physician and psychiatrist that she is, but as the grandmother that she may become, standing where I am now, perhaps remembering this moment when her son had looked at the magical book.

Roosevelt came and stood by me. Perhaps he could see that I was deep in reflection. We are not ones to indulge in public displays of affection, but I leaned ever so slightly against him, grounded once more in the miraculous present.

I stood again with my uncle Chow Woo at the altar, where we bowed and burned incense in front of the relics and images of our ancestors. This was all part of a pattern with which I was familiar, I knew the drill. But then it came time for something quite out of the ordinary. We entered another large outdoor space, which I had not seen when I first visited. Ten banquet tables had been set up for a festive lunch. We were all so new to each other that shyness descended on our group. But I was not going to let shyness affect how the afternoon unfolded. The purpose of this trip was not for the Americans to sit with the Americans and the Chinese to sit with the Chinese. I was committed to having people mix it up.

At the same time, I had to deal with certain linguistic realities. Of the 150 of us, perhaps ten were bilingual, so there could be one bilingual relative at each of the ten tables. Once they were situated, I moved other people around, persuaded them to sit in different configurations, moved cousins to sit near other cousins or near aunts and uncles. Then the feast began. We sat and started eating the mounds of fabulous food, but there was very little talking. I heard some bursts of quiet conversation from several tables, then awkward silence. I glanced over at Roosevelt, who was focused on his plate of noodles.

I am not sure who was responsible for it, but all of a sudden small glasses appeared and Moutai was poured into them. Moutai is often described as Chinese wine, but in fact it is a high-proof—about 65 percent—very intense whiskey made from sorghum. In the 1950s it was the official drink of the People's Republic of China, and at state dinners or meals with visiting dignitaries, Moutai is poured and also distributed as a gift. Now, those of us who are Jamaican practically grew up drinking J. Wray & Nephew overproof rum and are no strangers to the power of high-proof alcohol in changing moods and loosening up social dynamics.

I was impressed at the change after we shared the Moutai. The next thing I knew, the volume went up. The clatter of conversation and laughter ricocheted from table to table. When our Chinese relatives saw that we liked it, we somehow passed an important test: we were family and we were accepted. The lunch became one big, convivial party, just short of a bacchanal.

People got up and started wandering from table to table. Who is this? Whose brother/cousin/uncle/aunt/sister are you? Whose brother/cousin/uncle/aunt/sister am I? People dispensed with translators and used the Esperanto of pantomime, pointing, and nodding yes or no. Instead of silence, there was the sound of laughter and happy conversation. At one table, I saw Roosevelt with my first cousin Minzhang, the son of Chow Ying, one of Samuel Lowe's sons who had died. They had their arms around each other and were toasting each other with glasses of Moutai.

It was all Roosevelt needed. I watched him work the room, making an appearance at every table. He went from his contemporaries to the elders, toasting each one individually. Which is to say that by the end of the meal, Roosevelt had clinked his glass about 149 times. The Chinese have no word for cousin, so we taught them to say "cousin" and it became our family toast. No matter what the relation, when a glass clinked, the word "cousin" was spoken, pronounced as variously as possible for just two syllables. "CuhZIN," someone would say. "COHsyn!" others would chime in. And in the middle of it all was Roosevelt, "Cousining" anyone whom he came across.

My husband has a great big booming barrel laugh and I heard it from all corners of the courtyard. And when he arrived to clink glasses with Aunt Adassa, she smiled with such delight and sweetness that she looked like a young woman again. She loved him. I basked in the sheer, glorious fulfillment of it all. A few toasts were offered and then it came time for me to go to the microphone.

I have given many speeches in my life and this was probably not the most eloquent. I looked around the courtyard—at the mosaic of brown and yellow faces; at small children and babies and young adults and teenagers and middle-aged men and women and my elderly uncles and aunts; at people bound together for generations by a man who had the audacity to travel from China to Jamaica over a hundred years before. And now 150 of us were here, and many more were planning to come for Aunt Adassa's birthday party. The wall at the village of Lowe Shui Hap came alive in a new way at that banquet.

When I stood up, I asked for the youngest members of each family group to join me. The youngest of Samuel Lowe's grandchildren came up to speak. And then I asked for the oldest members of our family to be recognized. The remaining children of Samuel Lowe sat, beaming. I toasted them all, and we all toasted one another. I looked over at Roosevelt. He raised his glass to me in a silent toast. And I knew that my proud Black husband had found his Chinese family, too.

WHERE MY ANCESTORS
ARE LAID TO REST

THE NEXT MORNING WE BOARDED TWO BUSES FOR THE NINETY-MINUTE DRIVE to Guangzhou, the cemetery where our grandfather, his wife, and their eldest son, Chow Ying, are laid to rest. The most important grave for me was my grandfather's. The bones of the founder of our clan, who started the building of the village Lowe Shui Hap in the late 1700s, were buried somewhere in the province but not at this cemetery. At one point, his bones were kept near the village, but they were stolen by bandits and held for ransom. The bones were so precious that the remaining family members paid whatever was asked. After this transaction was repeated three times, the family decided to move the bones far enough away to make it impossible for thieves to take them.

The cemetery is a vast area, dense with headstones of elaborately carved marble; pictures of the deceased are set into the tombstones, above golden Chinese characters. We were there to pay our respects to our grandfather, but only one of our elders—the youngest, Aunt Anita, who now lives part-time in Australia—had come with us. Elders and babies generally do not go to cemeteries, since it is thought that the energy of the dead and the energy of those two phases of life are in stark disharmony and should not be mingled.

I have also always felt adamantly that children do not belong at a funeral. When Imani, my daughter, was nine months old, her biological paternal grandfather died and her biological father told me that she and I both needed to attend the funeral. I told him that I was not going to do that. "I hope that your family understands," I said. "But I am not taking her to the funeral."

"Oh, but you have to," he replied, astonished at what I was saying.

"No, I am not going to," I said. "Babies don't belong at funerals. I will never do that. It has nothing to do with your father, but your father is gone. I am not taking my child to a space where there is such a concentration of grief and sorrow and death."

I am not sure that he ever truly understood, but in China, when I was told that the Chinese don't take babies to funerals or cemeteries, and that the elderly also don't go to these places because they are too close to the time of their own passing, I understood where my own feeling must have originated.

Slowly, we proceeded through the field of marble until at last we arrived where my grandfather was laid to rest. Solemn, respectful, preoccupied with our own individual thoughts, we lit the incense; only the sound of matches striking the flint could be heard. I stood with my family, the smoke swirling gently around our faces, and bowed and prayed over his grave, and the graves of his wife, Swee Yin, and Uncle Chow Woo and Aunt Shizhen Xiao's firstborn son, Min Kai.

As we slowly descended the steps that led us to the path back to the bus, I was overwhelmed by grief that I had never before experienced, not even when my mother died in 2006 or when my father died three months before. I was ambushed by a profound, almost existential sense of loss and began to sob uncontrollably. (And, I might add, uncharacteristically.) I suppose that I could explain the feeling as having been triggered by a cathartic release of long-suppressed anxiety about this trip. Or maybe it was just fatigue and jet lag. Or maybe some inexplicable emotional chord within me was struck.

Whatever the reason, I couldn't even keep walking. I stopped on the steps, leaned against the retaining wall, put my head in my hands, and sobbed. My cousin Andrea—Gilbert Lowe's daughter who lived in Brooklyn—took me into her arms. We had gotten to know each other only over the last few months, and yet she was as comforting to me as a sister. She murmured soft words that I cannot recall, and held me for a few minutes so I could compose myself.

I felt drained by this experience at the cemetery. I had talked to my grandfather before, and I had listened to him as best I could. But nothing had prepared me for what it would feel like to fulfill this dream. Being by my grandfather's grave was the closest I had ever come to him physically. It was also the closest my mother had come to him since she was three years old. It was as if her presence within me had been activated at the grave, and it was all too much.

I took a few deep breaths, wiped my eyes, blew my nose, and gave Andrea a grateful hug. As we slowly walked back to the bus, Andrea spoke about her father, Gilbert; and I spoke about my mother, Nell.

"My dad was always so sad," said Andrea, who was still coping with the pain and the uncertainties of growing up with a depressed parent.

"Yes!" I said. "If there were an overriding description of my mother, I would describe her as fierce—God, she was fierce—and sad."

This profound sense of sadness that Nell and Gilbert lived with cast a pall over their entire lives, their entire beings. For Andrea and me, our parents' melancholy was a defining characteristic. They shared a father; they shared DNA; and now I know, they shared abandonment.

I'm haunted by my mother's refrain, "You don't know what it's like to grow up without the love of a father." Gilbert's melancholy mimicked Nell's. And they both struggled with the reality that their father left them behind. Samuel Lowe wanted his children with him but in the end, they were abandoned.

Nell's mother abandoned her again after taking her from Samuel, who did not intentionally leave her. Gilbert was abandoned in Mocho; yes, with his mother, Emma Allison, but he knew he had sisters and brothers and a stepmother, too. They all left for China and left him behind.

I've begun thinking of this in another way—as an African-American descendant of African slaves. When my African ancestors were stolen from their families, their lands, their culture, what profound sadness and despair immediately overtook them?

Shackles, filth, beatings, rapes, murders. Europeans told them they were not human, had no souls, had false gods. In just one generation, my mother and uncle suffered lifelong depression because their identities were damaged when they were separated and abandoned.

Would kidnapped Africans have felt abandoned, not rescued? What pathologies are imprinted on us, the descendants of African slaves? And, still we survive. Some of us have hidden scars and others identifiable scars—but the barbaric ravages of slavery have been indelibly imprinted on us for at least three hundred years. I think of Dr. Maya Angelou's chant for survival, "And Still I Rise."

Both Gilbert and Nell had been severed from their roots, and that cannot happen without terrible consequences. Even as adults with their own families, such people walk through life feeling incomplete. My mother always felt that way. I realized, from Andrea's comment, that Gilbert did, too. The moment at the cemetery changed everything. I was, at last, complete.

AUNT ADASSA
TURNS NINETY-FOUR

THE GREAT CELEBRATION FOR AUNT ADASSA TOOK PLACE THE DAY WE WENT to the cemetery. There were even more of us at this party than had attended the gathering the day before. Little children were hugging each other. A big family tree that my first cousin Man Kwan, Uncle Chow Woo's son, had created was posted, and family members pointed out various names and locations. There were pink balloons and ribbons and an enormous card for us to sign. There was also, of course, Moutai, poured in great quantities to lubricate the conversation. Aunt Adassa sat happily, wearing her little woolen cap, accepting the love and the acknowledgments of all who came.

After a series of toasts, Man Kwan stood up and asked for the attention of the assembled guests. His beautiful only child, his twenty-five-year-old daughter Siqi Luo, was by his side, translating his words into English. I had thought that he was going to say something more about Aunt Adassa, but I was wrong. He began by noting that this was "the most united reunion of the Lowe family in almost one hundred years." There was a brief interval of applause. "I would like to thank Paula for all the documents and information she provided, without which this reunion would not be possible."

I felt a little embarrassed but got up and took a bow, as the applause rained down. My cousin explained that the genealogy book in Lowe Shui Hap, though impressive, was basically flawed because during the Cultural Revolution—when everyone was suspect, when one relative who might be perceived as unreliable could jeopardize the safety of a whole family—the Lowes had stopped making entries in the book. "Now we regret that some parts are missing," he said, looking apologetically at me. "I am Samuel Lowe's grandson. I have a responsibility to my grandfather. I am duty bound to complete this book and to put it in good order."

Nell and her children and their children will be on the wall.

Gilbert and his children and their children will be on the wall.

Our family will be united from now on. The chain that had been broken was reconnected at last.

On our final night in Shenzhen we had our last large family party. The waiters wore Santa Claus hats, and Chinese members of the family sported the LA Sparks T-shirts that we had brought them as gifts. The theme could have been "Long live the blood of Samuel Lowe." My brother Howard stood in front of the group with eleven members of his family—his daughters and their children. My brother Elrick stood up with his son Chan, the only male surnamed Williams of his generation. I stood with Imani; my husband, Roosevelt, whose birthday it was; and my grandson Idris. Then it was time for pictures, and crazy configurations of cousins, aunts, uncles, grandparents, and children assembled themselves for the camera.

My uncle Chow Woo told me that we all had to come and visit again. "Bring everyone from Jamaica!" he said. He pointed out that our mother wasn't originally included in the seven siblings. "But now we include her among the eight siblings!" he said proudly. My uncle took in the room and announced, "Now there are three hundred and twenty of us, including Paula's family. It makes us feel very proud. Other families cannot compete with our family!"

The family motto: "Prosperity, Family, and Education."

But above all else, on this special night, family.

THE RETURN
OF THE JIA PU

I HAVE LANDED IN CHINA MANY TIMES, AS A BUSINESS EXECUTIVE AND AS A vacationing tourist. But recently, going to China has given me a reassuring sense of coming home. It wasn't just that I knew where I was going, or how to maneuver in the demanding Chinese culture, I felt as if I *belonged*. I was not merely a tourist doing some retail therapy and adventurous eating. I was someone who had a place and a family. My heritage was in China.

But in the year since we initially met, my uncles and aunts were becoming even more frail. When I first arrived in China, my uncle Chow Woo would deftly maneuver his chopsticks on a serving dish; snag a helping of chicken, long beans, or bok choy; smile at me; and deposit it on my plate. This was not simply a gesture of hospitality. Manners and hierarchy are important in China, and an elder serving a younger person is conferring a great honor and signifying love, regard, and respect. Accordingly, in 2012, Uncle Chow Woo, who was in his mid-eighties, served food to the sixty-year-old daughter of a sister he never knew existed.

Then, between December and August, my uncle developed some health challenges—he had kidney trouble, for example, and was hospitalized for weeks. It was sad to imagine that after all our time apart, cruel fate might rob us by taking him away. My cousin Siqi would send bulletins of his health to me, and I would share them with the clan of Black Chinese Jamaican Lowes in the United States, Canada, and Jamaica.

Still, Uncle Chow Woo was clearly not ready to go. He was still attentive and engaged. On my return to China in August 2013, I sat next to Uncle Chow Woo and it was my turn to show love and respect by serving him. "Would you like soup, Uncle?" I would ask in English. He would nod and make a throaty sound that he knew I would understand as an affirmative. How about some beef, Uncle? No, no beef, he'd indicate. Rice? Yes, always rice. A little French cabernet? But of course.

No one intervened. No one interpreted. Uncle Chow Woo was content. I was at peace. In my mind, I had a conversation with my grandfather and my mother. Had they been physically present, I would have served them, gently and lovingly placing the food on their plates. I listened to the harmonies of the Cantonese, English, and Hakka being spoken by the Lowes and my traveling companions—a contrast to the cacophony of the marketplace. The sound of all those languages was soothing, and brought me an even deeper sense of peace. I looked at my uncle and he caught my eye. We understood each other without language. Instinctively we each knew what the other was saying even though we did not have a word in common.

I looked around the table at all the history that was represented there. We were all sharing this voyage of discovery of our family. Having an animated conversation with my friend Marcia was my cousin Wei Yi, Aunt Barbara Hyacinth's eldest son. He had been born in China, and one day he asked his grandmother Swee Yin—who by then was a widow—why no Lowes lived overseas. We are all educated, he said. Why hadn't anyone been dispatched on a foreign service assignment for China?

"My grandmother asked me in a very quiet voice, 'Don't you know your mum was born in Jamaica?' " he said. He was stunned.

His grandmother led him into her bedroom and began to rummage under her bed. She dragged out a box tied up in a cloth. Almost conspiratorially, she gingerly unwrapped the box and extracted the contents. I thought of my mother and the box hidden under her bed. A box full of treasures under a bed seems to be a point of unity among Chinese women in our family.

Wei Yi watched as his grandmother pulled out six Jamaican passports that had been issued to Aunt Adassa, Uncle Chow Ying, Uncle Chow Woo, Uncle Chow Kong, Aunt Barbara Hyacinth, and Aunt Anita Maria. They had the royal blue Jamaican cover, and when Wei Yi opened them, they appeared to be brand new even though they had been issued in 1927 and 1933.

"I could not believe it," he said. "No one in the family had ever talked about living in Jamaica." Of course his children had known that our grandfather had spent time in Jamaica, but that, and the few English words he had taught them, was the extent of it. During the Cultural Revolution, any time spent abroad was a cause for suspicion. Many documents, records, and family chronicles were burned to destroy any traces of bourgeois, capitalistic values.

Similarly, such traditions as the generation name have all but ceased in today's China. The males of my generation have Man as their common name. All of the female children born in my generation have the family name, written as Lowe, Luo, Law, or Lo. The next name is Siu, meaning happiness or laughter. And the individual name comes last. My name is Lowe Siu Na.

Today, parents don't give a generation name. Historians are already finding it difficult to trace families because, for self-protection, during the Cultural Revolution most families destroyed their *jia pu,* or genealogy book. That destruction wiped out not just the past, but also the future. In the past, when all members of a generation shared a name, a family could date each person's lifetime. This tradition eventually was celebrated in the family's poem. The traditional poem of our family contains some forty generation names, in an easily memorized verse. Afterward the generation names are repeated over the next forty-year cycle. And the next, and the next. . . . In the same way that my grandfather taught my mother to count in Hakka, parents taught their children the family verse—one generation after another for hundreds, even thousands, of years.

During the catastrophe of the Cultural Revolution, these traditions were deemed not beautiful but bourgeois. Distinguishing one family from another was considered audacious and inimical to the solidarity of a classless society. In fact, such traditions were tantamount to a sign pointing out class enemies, who were considered dangerous because of the legacy of books and poems written for just one family in words that carried beauty and tradition across the millennia.

CHINESE NAMES

I WANTED TO HAVE A CHINESE NAME. AFTER MY FIRST ENCOUNTER WITH MY family in China, after my spiritual connection with Uncle Chow Woo and Aunt Adassa, I wanted a kind of baptism. I wanted something to signify that I had now entered and become part of another world. I was no longer a tourist in China; I had confirmed and embraced the Chinese part of my soul and it needed to find an outward expression.

When my grandfather went to Jamaica, his name was Lowe Ding Chow and looked like this: 罗定朝. But there was no pinyin or Wade-Giles, no standard way to romanize the Chinese characters. In the end, he became Samuel Lowe. I found myself longing to penetrate the world of Chinese characters, to understand the way each character symbolized the sound and meaning of a word. And I wanted a name of my own that would link me with my grandfather and my other Lowe relatives.

After I returned home, two days before my sixtieth birthday, in the Year of the Dragon, I asked my young cousin Gary, whose English is strong, to ask Uncle Chow Woo, on my behalf, if he would do me the honor of bestowing a Chinese name on me. I felt jittery and vulnerable in a way that I am not accustomed to. I worried that perhaps I was overstepping, and that this wish of mine would not be granted.

The next day, I received an e-mail from my cousin. He wrote on behalf of Uncle Chow Woo:

Paula, my grandpa has thought about a Hakka name (Chinese name) for you and your brothers. See if you like it.

As I mentioned before, your name will start with Siu and your brothers' name will start with Man, which is the same as the 3rd generation of Lowe family.

Your Chinese name would be Law Siu Na (罗笑娜), Na is a similar pronunciation to your Paula's la and it means beauty.

As to your brothers, Elrick was named Law Man Chi (罗敏志) and Howard was named Law Man Kin (罗敏坚). The Chi and Kin in Chinese represents your willing to reconnect with the Lowe family is strong. How do you think of these names?

I looked at my name: Law Siu Na. I rolled it around in my mind and said the three syllables out loud. It means "beauty," he had said. "Law" is Lowe. He had given me my Chinese name. I felt tears stinging my eyes as I tried to write a response that would come close to the emotional power of this experience.

"I'm moved to tears, Gary. Please thank our uncle so much for us," I wrote. "We will now have to learn to write our names! See you in December!"

I also wanted my mother to have her Chinese name. This was a bit more complex because in China people are not named posthumously. I worried that my uncle would see this request as a violation of tradition or, worse, would find it insulting.

Indeed, my uncle hesitated, but he finally decided that the circumstances were extraordinary and tradition could not be a reliable guide. This time my young cousin Siqi wrote the e-mail:

Chow Woo finalized on the name 羅碧珊 pinyin: luo bi shan, pronounced: law bik saan (Cantonese).

-碧 (bi) is the generational name for their generation; i.e., Adassa is 碧玉, Barbara is 碧珍, Anita Maria is 碧霞.

It means the color bluish-green and can mean jade.

-珊 (shan) means coral reefs, it reminds people of the beautiful island of Jamaica, representing Sam's longing for the daughter in Jamaica; it is also the closest character that sound like Sam, we know that in the west, the eldest son always inherits the father's name, so he wants Nell to have his name too.

On another note, 羅碧珊 is traditional Chinese writing, it looks more beautiful and I think Samuel can have his name written in traditional Chinese too (羅定朝). The simplified Chinese was introduced in 1956, this was way after he was born anyways. In fact the difference is only the Lowe character. This is just a suggestion though.

Could my mother's name have been any more beautiful? The coral reefs, the longing for the daughter in Jamaica, the character closest to the one that sounds like Sam—and that it is beautiful in Chinese writing. Of all possible names and images, my uncle had found the ones that evoked the sea and the island. Nothing could have been more perfect for my mother.

I remembered the day when we scattered my mother's ashes in the water off the coast at our home on Martha's Vineyard. My mother had died in 2006, at age eighty-seven, at a nursing home in Evanston, Illinois. She had still been beautiful, but her mind had gone. I was in Los Angeles; Howard was in Florida; but Elrick had settled in Chicago, and Evanston is the suburb of Chicago where he and two of Howard's daughters, Lynai and Leah, lived with their families. We had all agreed that my mother would live out her life there, surrounded by family.

Nell was very spiritual, connected with the universe. She had always felt that there is as much unseen as seen, in this world. She would talk about uncles, other relatives, and ghosts returning for visits. They would leave all sorts of traces of their presence: moving a photograph, appearing as an animal, showing up in a dream.

Her grandchildren would often say to her, teasingly but also with an element of truth, that after she died she should give them some sign that her soul had endured—that she had made the journey to the other side safely. "What kind of sign would you like?" she would ask. "How shall I appear?" One of Howard's daughter suggested that she return as a white bunny. My mother, who was much warmer and sweeter with her grandchildren than she ever was with us, agreed.

My brothers were in Evanston during the April days when our mother was dying, and the daily vigil of visits and waiting became part of our family life. One day near the end, Lori walked to the nursing home to visit Nell. It was a beautiful spring morning, sunny though still chilly—the kind of morning that, after a Chicago winter, holds some promise of milder days ahead. As Lori walked down a path bordered by green bushes, a white rabbit raced across it in front of her. She stopped and gasped, remembering her conversation with her grandma. The rabbit was part of the springtime peace and beauty; the rabbit was a sign. Still shaken, Lori took the elevator up to the fifth floor and went into her grandmother's sun-drenched corner room.

Nell had died just a few minutes before.

I know that there are coincidences. I know that rabbits can be everywhere (they are, after all, known for their fecundity). I know that a rational mind would see no connection between the white rabbit and my mother's death. But my mind is not rational about this. The rabbit was a sign from her. And I know that my discovery of her Chinese family was not simply a result of good fortune and hard work. Of course both luck and work mattered, but the luck had an element of intercession from the other side. Nell's presence accompanied my search for her father, and often led the way.

After she died, we decided to wait until the summer to scatter her ashes, since we planned to cast them into the Atlantic, off Martha's Vineyard, a place she didn't know but would have loved.

The children and the grandchildren all gathered there on a beautiful August day. We wore our bathing suits and cover-ups and went down to the beach with the urn. There were about twenty of us, ranging in age from one year old to my brother Elrick, now the oldest member of our family. I had asked the young family members the day before to find small items that reminded them of Nell. They came with beautiful seashells, flowers, herbs, even some writing. Together we went and prepared to say our final good-byes.

I opened the container, held her remains low near the water, and gently teased the ashes onto the waves. As her ashes floated on the water, the tide kept washing them toward the shore, not to the open sea. Soon all the young children leaped into the water, swirling her ashes and pushing them farther out into the sea. They made S's with their arms, working against the tide. "Go back to Jamaica, Grandma," they chanted. "You're free now, Grandma. You can go back home!"

I stood on the beach and said good-bye to my mother.

And yet, was it really good-bye? Is it ever *really* good-bye?

DREAMS OF
MY GRANDFATHER

I STILL DREAM OF MY GRANDFATHER, OF MY MOTHER, OF THE LIFE THEY BOTH could have led had they known about each other—and of the life I would have led had he not disappeared when my mother was three years old.

But my dreams are different now.

I can now summon my grandfather's face, and I can recite a biography that did not end in Jamaica. I know his children, his grandchildren, and his great-grandchildren; have seen his house; have walked into his store; have prayed at his grave.

Now when I say "Grandpa" out loud, it is not into an empty space where my voice sounds lonely and part of me feels pathetic. When I say "Grandpa" it is not as if I am trying out a new word and failing to get it right. Now I say it to others who also call him Grandpa. I refer to him as the subject of a sentence, or as the direct object of a verb. He has become thoroughly integrated into the warp and woof of my life, because he and the rest of our family have become a day-to-day, morning-to-evening, palpable, visible part of my world, as my brothers and I have become part of theirs.

Now the atmosphere that we grew up in, an atmosphere charged with a sense that something was desperately wrong, has at long last been put right. The fundamental contradiction—that family is the most important thing in the world, but that our family was so small and fragile—is gone. It has been replaced by a conviction that family is the most important thing in the world, period. No matter how long the family has been divided by circumstances, by oceans, by continents, by time, by historical cataclysms, by race, its importance endures. The connections transcend the divisions.

In looking back on this great journey that I have undertaken, I realize that as important as the destination was, as overwhelming as the discovery of my Chinese family turned out to be, I have also discovered soul-stirring meaning in the journey itself. The miraculous appearance of just the right people at just the right time; my cousin JJ insisting I attend the Hakka conference; Jeanette Kong and Keith Lowe leading me; the revelatory moments when my brother Howard counted in Hakka and my brother Elrick remembered the year of our grandfather's departure from Jamaica; my own anguish—beautiful in retrospect—at my grandfather's store and at his grave; my loneliness: all this has led to the moment of peace when I held hands with Aunt Adassa and the moment of joy when I caught Uncle Chow Woo's protective, proud look in my direction.

When I was a little girl, I would talk to my grandfather, even though I did not know much about him. I would ask him, "Where are you? Why did you leave?" Now, I talk to my grandfather as a man whose essence I have grasped, a man whose suffering I have felt, a man whose legacy I have embraced. And I say to him, "Thank you." Or, "Did you see that?" Or, "We are all at peace."

I had promised myself that I would find him—that the lost puzzle piece would be restored, that the interrupted story would be continued, that the broken lives would be repaired.

And I did.

And they were.

And the world as I know it is a sacred, miraculous place.

EPILOGUE

THE END OF ONE STORY BECOMES THE BEGINNING OF MANY OTHERS.
I could continue my search for Samuel Lowe, because even
though it has ended in some ways, new narratives have appeared,
and different currents have pulled me in new directions, always with
Samuel Lowe somehow being present. I could write about how, once I
found him, his presence in our lives changed my family. That change
is a new story. I could focus on how my conversations with my brothers are altered; I am learning from them and they are learning from
me. In the process of traveling and forming new relationships, we
three Williamses have been rethinking our own history, because that
history is dependent on our mother's story, and her story is no longer
what it once was.

Of course the outlines are the same. The woman who pressed
a meat cleaver against the throat of Fre-Zee's father will forever be
that woman. The woman who first married, then stabbed our father
will forever be our mother. The woman who protected us so instinctively and so passionately will always be that woman. But now we can
imagine what the woman who stared out the window might have been
looking for. We can hear her counting in Hakka, not as an anomalous
act by an unpredictable mother, but as a long tradition perpetuated by

a little girl sitting with our grandfather—both members of a family that has lasted for three thousand years. We can easily imagine her not alone but as a sister to siblings—who were always invisible to her, always out of her reach.

I could also write about my daughter Imani and the children of her generation, whose perception of how they fit into the world has changed. They are what they were before our trip to China: educated, well-traveled young adults some with children of their own. But now they know that their lives have a different purpose and significance, and that the World Wide Web means something different, personal, and even autobiographical to them. They know that their boundaries—their ideas of who they are—have shifted profoundly.

I could write about my Chinese family. My Hakka friend Jeanette Kong observed that we were blessed because of the way our family reacted when we appeared. She could imagine a very different response—a Chinese family dismayed, perhaps, by the presence of Black people in their lives; or angry at us for intruding on their peaceful, uncomplicated, prosperous existence; or merely cold and dismissive, hoping we would disappear. There were many possible responses, but the way the Lowes embraced us opened a future to us all.

Family. Prosperity. Education.

We are all entrepreneurs, and we are committed to our family's enduring presence in this world long after my brothers and I have died. We have had meetings over the year, with representatives from the families of each of Samuel Lowe's children, to discuss the family business. It's in our blood. We have created Lowe Family Enterprises, comprising Samuel's grandchildren from five nations. We are committed to expanding our grandfather's vision and to creating still more wealth and legacies for our family.

Then there is the family of Gilbert Lowe, my mother's half brother. Three of his daughters lived a subway ride away from us for years—not that any of us ever imagined such a possibility. With the appearance of Gilbert's family, his sister, Adassa, has become a source of revelations for us; and our arrival has opened, for her, a flood of memories that never before had she been able to share.

She remembers that when she was a little girl she went with her father to visit her mother, Emma Allison—they seem to have devised an almost modern version of joint custody—even after Samuel Lowe had married and had children with Swee Yin. Adassa recalls that she called Swee Yin "Madame," and that her Jamaican mother would always remain her mother. Aunt Adassa also remembers another boy, her older brother Aston Samuel, who died when he was only six. We had all supposed that he was a child of Emma's, born before Samuel entered her life, and we assumed that his surname was Allison. But Aunt Adassa had known the truth for over ninety years: Aston Samuel's name was Lowe, and he was the first son of our grandfather. The proof was his middle name, Samuel. Aston Samuel is buried in the front lawn of Emma's home, with his mother nearby.

What does this discovery of yet another child of our grandfather's mean for our family? Probably not much.

Does it matter? Probably not. Life went on. Nell and Gilbert stayed in Jamaica. Adassa and her father and the rest of the family went to China. And if a firstborn son died as a child, did that fact, that secret, affect the future?

Of course not. And of course yes—because Adassa's memory of this boy, and our discovery of him, coming only after we had discovered so many other bits of family history, is a reminder of the richness, mysteries, and imponderables that make up families.

We can go through life believing that what we know is a kind of truth. Not the basic kind of truth—guilty or innocent, journalistic, did the child put her hand in the cookie jar—but truth that is more existential. This existential truth defines us until something shifts a bit, a new puzzle piece appears, a stone is turned, a stranger e-mails us to say, "Samuel Lowe was my father."

And then the whole world changes, even if all our familiar landmarks remain the same. My brothers and I will keep working hard in our businesses, now with some Chinese cousins as partners. My grandson will keep playing with his grandpa—my husband, Roosevelt. My daughter, Imani, will see patients; Elrick's son Chan will continue to be the only male in his generation with the surname Williams; and Howard's daughters will continue to work hard and care for their families. We will mourn our father and our mother. And eventually we will mourn our elders whom we have been blessed to know and love.

Los Angeles, 2014

ACKNOWLEDGMENTS

My successful search to find my mother's family would not have been possible without the immense help from my father's family. My father's siblings and cousins provided the first meaningful clues: my aunt Carmen "Cutie" Velez contacted our cousin John "JJ" Hall in Toronto, who connected me to the Hakka-Chinese-Jamaican community in Toronto. Simultaneously, my aunt Ouida Harrison in Kingston connected me via e-mail to Eily Chin Lowe, who introduced me to her cousin-in-law Raymond Lodenquai in Toronto.

From JJ's connection I met Jeanette Kong, who would become not only the director-producer of my documentary *Finding Samuel Lowe: From Harlem to China*, but my dear Hakka-Chinese-Jamaican sister-friend. In June 2012, Jeanette and I met at the Fourth Toronto Hakka Conference, cochaired by Dr. Keith Lowe and Carol Wong, who both became very important to me and to my search.

So many helpful Hakka people were at the conference, and they committed themselves to trying to unravel this mystery. I met Keith's sister, Barbara Lowe Eckel, who lives in Atlanta. She gave me a brochure she'd brought from the largest Hakka Cultural museum in China. She had visited the museum that was her family's ancestral village. The village?—Lowe Shui Hap, which turned out to be my own grandfather's birthplace.

Jeanette Kong, who researched many of the documents and records in Jamaica and in the United States, spent days online, on e-mail, and on the phone sourcing Lowe family documents from decades ago. It was Jeanette's documentary, *The Chiney Shop*, about the Chinese shopkeepers in Jamaica that spurred me to beg her to help me find my family and also to serve as producer-director of my own documentary. I owe her boundless gratitude for her work and her friendship.

After Jeanette's cajoling, Keith queried his family in China, and within twenty-four hours, I'd found my grandfather's descendants. I will always love him, and I can't thank him, his beautiful wife, Amoy, and their sons enough.

My former NBCUniversal and GE colleague and stylish sister-friend Marcia Haynes and I were together in China in August 2012 when I met the first members of my Lowe family. E-mails from Keith Lowe's nephew, Yiu Hung Law, set up that meeting. Thanks to my beautiful and dear friend and her camera, I have photographs from that first meeting. And thanks to cousin Yiu Hung for sending Keith's e-mail to Uncle Chow Woo. Yiu Hung's e-mail changed my entire existence.

My young second cousin Siqi Luo deserves special mention because the intelligence, patience, and dedication she has shown to our family's efforts to unite have been unmatched. Siqi is the connector between the Lowes in the West and the Luos in the East. That leads me to explain a bit about our name. This is the Chinese character for our name: 罗.

As the name was written in various languages using the Western alphabet, it is written Lowe, Luo, Lo, Law, and Lau, yet they all are the name 罗. Samuel Lowe's Jamaica-born children spell their surname "Lowe," yet their children born in China spell their surname "Luo."

With that preface, I'd also like to give special thanks to Siqi's grandfather, my uncle Chow Woo Lowe; her father, my first cousin Minjun Luo; and to my aunt Adassa Lowe for being so loving and embracing. You will greatly be missed. (Aunt Adassa passed away during the production of this book.) Minjun has updated and corrected our family's *jia pu*, or family genealogy book. After meeting us, Minjun revised our grandfather's history to include his two children, who were not taken to China.

I want to thank my beautiful, wonderful friend Marianne Szegedy-Maszak, my comrade from when we both joined the board of the Center for Public Integrity nearly two decades ago, and who served as a brilliant taskmaster for my book. Marianne is an award-winning journalist and author who in 2013 published an unforgettable and haunting memoir of her own family, *I Kiss Your Hands Many Times: Hearts, Souls, and Wars in Hungary*. Marianne was invaluable to me in the writing of my own book. Her thoughts, guidance, organization, and writing helped me formulate and, frankly, finish *Finding Samuel Lowe*.

Thank you to Tracy Sherrod, my sister-friend and senior editor at HarperCollins, who guided me, as a first-time author, successfully around many daunting challenges. And thanks to my friend, fellow member of the National Association of Black Journalists Doug Lyons, for introducing me to Tracy.

And to my loving and patient husband, Roosevelt: I can only thank God and my ancestors that he is in my life, that he is my soul mate. The inspiration I get from watching him and our grandson, Idris Morales, love each other is without question what motivated me to search in earnest for my own grandfather. Our daughter, Dr. Imani Jehan Walker, reminds me so much of my demanding, intolerant, and perfectionist mother that, even in my moments of exasperation, I love her and can't imagine my life without her. I dedicate my love and my life to them.

To my Gilbert Lowe cousins, thank you for your love, friendship, and support. Tony was with us on our first journey to find our grandfather's beginnings in Jamaica. Loraine and Andrea joined the December 2012 reunion!

And to my nieces, nephews, grandnieces, and grandnephews, my daughter and grandson, I give thanks to you and gain my inspiration from you. Lynai, Leah, Imani, Chan, Imara, Alexis, Ishmael, Carlyn, Idris, Sierra, and Andre Jr.—you all went to China to meet your family. You are the descendants of Samuel Lowe—the Honorable Lowe Ding Chow. You have a great legacy to continue, great values to uphold. Your grandmother and great-grandmother was a formidable woman, and she left you equipped to make her and your family proud. Live up to the Lowe/Luo name.

And so, these are the people I want to thank. They have inspired, helped, and, in many cases, willed *Finding Samuel Lowe: China, Jamaica, Harlem* into a reality:

Elrick Williams Sr.
Nell Vera Lowe
  Williams
Roosevelt Madison
  Sr.
Florence Madison
Elrick Williams
Chan Williams
Howard Williams
Alexis Rodriguez
Lynai Williams
  Jones
Carlton Jones
Imara Jones
Ishmael Jones
Carlyn Jones
Sierra Jones

Leah Williams
  Haskett
Andre Haskett
Andre Haskett Jr.
Phil Johnson
———————
Anthony (Harry)
  Harrison
Paul Harrison
Maxime Harrison
Enid Anderson
Glaise Anderson
Ian Anderson
———————
Harold (Charlie)
  Meade
Norman Davis

Beverley Davis
Gilbert Lowe
Loraine Elaine
  Lowe
Anthony Lloyd
  Lowe
Nesta Lowe
Andel Emile Lowe
Annie Marie Lowe
Glenford St. Joseph
  Lowe
Donald Barrington
  Lowe
Shalane Allysia
  Lowe
Audrey Colleen
  Lowe

Sascha Lee
 McDermott
Russell Andre
 McDermott
Judy Ann Lowe
Emile Andrew Lyn
Carol Shawn Lowe
Ruth Mary Lowe
Phillip Douglas
 Richards
Samantha Hillary
 May Richards
Andrea Lowe

Ellen Richards
 Lennon
Michelle Lawson

G. Raymond
 Chang
The Toronto Hakka
 Conference

Stanley Law
Peggy Lowe
 Young
Arthur G. Lowe
Winston Lowe
Granville Lowe

Patrick Lee
Loraine Lee

Stephen Young-
 Chin
Tsung Tsin
 Ontario

Dalton Yap
Vincent Chang
Marcia Harford
Robert Hew
Ray Chen
The Chinese
 Benevolent
 Association of
 Jamaica

Diane Houslin
Lauren Tobin
Debra Langford
Laarni Dacanay
Maria Huerta
Deborah Elam
Art Harper
Linda Richardson
 Harper
Lloyd Trotter
Teri Trotter
Janine Uzzell
Alex Canfor-
 Dumas
The GE African
 American Forum

John Doychich
Tonya Thomas
Alesia Magee
Williams Group
 Holdings LLC

Karl Rodney
Fay Rodney
The NY Carib
 News
Warrington Hudlin
The Black
 Filmmaker
 Foundation

Lee Gaither
Fred Paccone
Martin Proctor
The Africa Channel

LaFleur Paysour
Pat Jordan
Jack Jordan
Donna Knight
Sherry Bellamy
Sherry Sherrell
Yolanda Sabio
Maritza Myers
Mitzi Wilson
Charmaine
 Jefferson
Barbara Walters

Kiese Laymon
The African
   American
   Alumnae/i of
   Vassar College
The Cardinal
   Spellman High
   School Foundation
Delta Sigma Theta
   Sorority, Inc.
The California
   African American
   Museum

Carlton Smith, Irie
   Routes Jamaica
   Tours

Seth George
   Ramocan
Basil Lee
June Ngui
Sherwin Tony
   Chong
Winsland Williams
Alphie Mullings
   Aikens
Aida Yohannes
Dorothy Kew
David Priever
Gina Paige
AfricanAncestry.
   com
Ancestry.com
Familysearch.org

(The Church of
   Jesus Christ of
   Latter-day Saints)
The Jamaica
   Gleaner
The Museum of
   the Chinese in
   America
The National
   Association of
   Black Journalists
Chinese Jamaicans
   Facebook
   community
The National
   Archives

And finally, the family of Samuel Lowe:

Emma Allison
Adassa Lowe
   罗碧玉
Zhangping Liu
   刘章屏
Xinyue Liu 刘新月
Guanzhan Liu
   刘冠沾
Meiling Liang
   梁美玲
Meifang Liu 刘梅芳

Qiwen Liu 刘绮文
Jianwen Zhou
   周健文
Jiawen Xian 冼嘉文
Zixuan Zhou
   周子轩
Haobo Xian 冼浩波
Qiuyue Liu 刘秋月
Guoheng He
   何国衡
Huifang He 何慧芳

Tingting Ye 叶婷婷
Weile He 何伟乐
Yongjie He 何永杰
Junlin He 何俊霖
Junying He 何俊颖
Kim Yuet Lau
   刘金月
Ching Hung Mok
   莫澄鸿
Yeuk Nam Mok
   莫若岚

LaiSze Wong 黄丽诗

Kiu Yan Mok 莫桥茵

Chi Man Lau 刘志文

Kam Heung Hsu 许锦香

Hiu Man Lau 刘晓敏

Ka Leung Lui 吕家良

Cheuk Wing Lui 吕卓颖

Yongwen Liu 刘勇文

Shaohua Li 李少华

Xiaohui Liu 刘晓辉

Xiaojun Chen 陈晓君

Swee Yin Ho 何瑞英

Chow Ying Lowe 罗早英

Xing Liang 梁兴

Xiaoyuan Luo 罗笑源

Zhigang Li 黎志刚

Jianping Li 黎健萍

Kailiang Chen 陈恺亮

Minzhang Luo 罗敏章

Huizhen Wu 吴惠珍

Zhaokang Luo 罗兆康

Chow Woo Lowe 罗早舞

Shizhen Xiao 肖仕珍

Xiaoliu Luo 罗笑柳

Xijiang Liao 廖西江

Jie Liao 廖杰

Lisi Wang 王丽斯

Ziyou Liao 廖梓悠

Minkai Luo 罗敏凯

Jianmei He 何间美

Xuhui Luo 罗旭辉

Xiangqing Ma 马向青

Zhaoen Luo 罗昭恩

Minjun Luo 罗敏军

Bin Ren 任斌

Siqi Luo 罗思其

Xiaoling Luo 罗笑玲

Xiaoquan Wen 温小泉

Yuxin Wen 温玉欣

Jiaxin Chen 陈嘉鑫

Chen Yat Fung, Alfred 陈逸峯

Minsheng Luo 罗敏生

Limei Tan 谭丽梅

Ruizhi Luo 罗睿智

Bingna Lin 林冰娜

Chow Kong Luo 罗早刚

Yudi Lu 卢玉娣

Xiaofang Luo 罗笑芳

Keqin Li 李可勤

Wanwei Li 李婉维

Minqing Luo 罗敏庆

Haiying Deng 邓海英

ABOUT THE AUTHOR

PAULA WILLIAMS MADISON is Chairman and CEO of Madison Media Management LLC, a division of Williams Group Holdings LLC, a Chicago-based investment company. She spent twenty-two years with NBC and was most recently the Executive Vice President of Diversity as well as a Vice President of the General Electric Company. Honored for corporate leadership and community outreach, Madison was named one of the "75 Most Powerful African Americans in Corporate America" by *Black Enterprise Magazine* in 2005 and was included in *Ebony* magazine 's "Power 100." She is Vice President of the Los Angeles Police Commission. On the occasion of the 2013 Centennial Anniversay of Delta Sigma Theta Sorority, Madison was inducted as a member of Honorary Centennial Six, a great career achievement. A native of Harlem, Paula and her husband reside in Los Angeles. This is her first book.